pudoc

UNIFYING CONCEPTS IN ECOLOGY

Report of the plenary sessions of the First international congress of ecology, The Hague, the Netherlands, September 8–14, 1974

Editors: W. H. van Dobben and R. H. Lowe-McConnell

Springer-Science+Business Media, B.V.

1975

ISBN 978-94-010-1956-9 ISBN 978-94-010-1954-5 (eBook)
DOI 10.1007/978-94-010-1954-5

Originally published by Dr W. Junk B. V. Publishers in 1975

Contents

Preface

The complexity of ecosystems forms perhaps the greatest challenge for natural science. Even the first step to comprehensive analysis, namely a survey of the participating species, often forms a major obstacle. This makes it understandable that ecologists try to abstract general principles from the interrelationships of the multitude of species for use in their efforts to investigate ecosystem dynamics. Such 'unifying concepts' were the main theme of the 'First International Congress of Ecology' organized in The Hague in September 1974 by the International Association of Ecology (INTECOL), under the auspices of the Division of Environmental Biology of the International Union of Biological Sciences (IUBS).

This book contains the papers presented at the plenary sessions of the Congress and a summary of the discussions engendered by them. At the Congress over 800 ecologists from many countries, representing diverse disciplines such as limnology, botany, zoology, microbiology, agriculture, met together for a week. The study of ecosystem dynamics depends on mutual understanding and close cooperation, and to stimulate an integrated approach a number of main speakers were invited to contribute papers on notions such as energy flow, productivity, diversity, stability and maturity from different points of view. These invited papers were presented at the morning plenary sessions, followed by discussions.

Papers delivered in afternoon sessions: on investigations relevant to the main themes, on the general results from the numerous International Biological Programme research projects, and in short symposia on special branches of ecological research, were made available for consideration at the Congress as the 'Proceedings of the First International Congress of Ecology: structure, functioning and management of ecosystems'*.

To avoid misunderstandings it should be made clear at the outset that concepts are not falsifiable theories. The criterion of a concept is merely that it should be *useful*. A concept may be useful even when it is not possible to relate it to quantifiable properties of natural ecosystems except in a general way. For

* Containing 98 abstracts and papers and author index, 422 pages; published by Pudoc, Wageningen, The Netherlands, 1974.

instance the trophic level concept, however shaky in this respect, has nevertheless deepened our insight into the gross structure of ecosystems.

The usefulness of concepts such as productivity and energy flow seems clear enough, and notions such as primary productivity and secondary productivity can be used in abstraction for the separate species involved. Both productivity and energy flow have the advantage that their definitions are not necessarily controversial. This certainly cannot be said of concepts such as diversity, stability and maturity. The demand for clarification of these terms is not purely academic as they have been used in the guidance of management and conservation before being properly defined and investigated.

During the Congress participants agreed that the term stability can have several meanings. A system can be called stable when few changes are observed. When it shows regular fluctuations we can speak of cyclic stability. Both cases may be a matter of internal stability. Another clearly different criterion is whether a system is vulnerable to disturbances from outside. We call it stable if it does not react to definite interference, or when it returns to the initial state after a reaction. For the faculty to restore itself quickly the terms resilience and elasticity are in use. When the restoration takes a very long time we can attribute a 'global stability' to the system.

In defining the term diversity there is also room for different approaches. Is diversity determined simply by the numbers of species per unit area, as suggested by diversity indices? Or is it related to the variety of life forms (such as phanerophytes, cryptophytes and geophytes) in the vegetation? A tropical rain forest may be richer in plant species but poorer in life forms than a savanna. Diversity could also be used as a measure of the number or intensity of interrelationships between species within a community. In climax vegetation such interrelationships are generally better developed than in a pioneer community, whatever the number of species present. It is necessary to keep these problems of definition in mind when considering relationships between stability as some measure of resistance against disturbances and diversity as some measure of complexity.

Another concept often mentioned during Congress sessions was succession. This term is generally used for the sequence in time leading from a pioneer community through a series of developmental stages to a more or less stable system, designated 'mature' by Margalef. This development is generally accompanied by an increase in some kind of diversity. This phenomenon has led to the 'naïve, if well-intentioned, view that complexity begets stability, and its accompanying moral that we should preserve, or even create, complex systems as buffers against man's importunities' (May, p 164 this volume). However, May argues that the reverse is more plausible: a stable environment is a condition for a complex and delicately balanced ecosystem to develop and maintain itself. This is in agreement with the experience that a diverse community in a nature reserve can only be maintained when management, for example a long-established agricultural practice, remains unchanged, in other words when stability is secured. This does not exclude diversity in its turn from contributing to stability, especially to the 'internal' aspect. Diversity may impede the establishment of immigrant species, or the spread of a disease.

Mathematical models devised by May suggest that as a system becomes more complex it becomes more dynamically fragile. It is indeed not logical to suppose that a complicated machine is less vulnerable than a simple one. An ecosystem is, however, no machine. Evolution provides a very essential difference. During the development of a system the participating organisms are exposed to selection pressures. They can adapt themselves to special conditions, which parts of a machine can never do. This very important faculty is not taken into account in mathematical models. Another important factor which cannot be neglected is microenvironmental heterogeneity; this gives protection against the disappearance of a species from a system during adverse periods, and thereby contributes to stability.

It is logical, furthermore, that systems in the course of succession attain a form of stability, whether they become diverse or remain simple. Stability by definition persists. Every unstable relationship will be replaced by another until by evolution of the species present or the establishment of new organisms in the system, some form of stability is reached, perhaps by extinction. This does not, of course, explain stability. It only means that in a living community stability is generally possible, and will be realized at some time.

A comparison with human techniques is relevant here. About a century ago the principle of the motor car was developed, but if at that time an inventor had presented a blueprint of the modern version of the combustion engine it would have been received with scepticism. It would have been considered too complicated to be reliable. In fact, its proven reliability is the result of a long process of trial and error. Generations of technicians have replaced parts of the engine with changed versions and added 'sub-systems' until reliability was acquired. These activities can be compared with the processess of succession and evolution in natural communities. It is not diversity *per se* which generates stability, but only a form of diversity aquired along lines of development with stability as a criterion. However, a system evolved in this way can be expected to be stable only with respect to the disturbances encountered during this development. A motor car is resistant to shocks caused by driving on a bad road, but not against sabotage by the addition of acid or sugar to the petrol. In the same way a rain forest may be more resistant to disturbances caused by a tropical storm than to those caused by the construction of a road through it (as observed by May, p 167).

Discussions on ecosystem management formed an essential part of the congress. The science of ecology is certainly not yet in a position to give exact rules for every management problem. There is, however, enough knowledge to warn against the exploitation of natural resources without a sound evaluation of the consequences.

The oldest form of ecosystem management is agriculture, beginning purely empirically by trial and error. As Holling observed (p. 248) small-scale errors in management lead to small failures, but large-scale errors can be disastrous. There are many historical examples of the collapse of agricultural systems due to mismanagement, especially in the arid zone which is vulnerable to primitive techniques such as burning and overgrazing.

Until recently large areas of tropical rain forest have escaped destruction, especially in less populated countries. Now that modern techniques enable man to exploit this biome on a large scale, the greatest care is needed to avoid adding another major failure here to the other historical examples.

We sincerely hope that the results of the First International Congress of Ecology will emphasize that the maintenance of vulnerable natural resources demands long-term policies with a sound scientific basis, and that neglect of ecological rules for the sake of immediate profit spells disaster.

THE EDITORS

Session 1

Flow of energy and matter between trophic levels

Chairman: E. P. Odum

Diversity as a function of energy flow

Eugene P. Odum

Introduction

Ecologists, in common with many other professionals, have their *articles of faith*, that is, oversimplified and generalized beliefs which may not stand up under deductive analysis. One such belief is that high diversity is a property of stable, natural systems and, accordingly, is a desirable feature for the systems of man. A comparable article of faith in economics is the idea of the *economy of scale* or bigger is better with unit costs declining with output. In reality, relationships between input quantity and output benefit, when plotted as *performance curves* are not ever-rising curves but convex or 'humped-back' curves. Which is to say that too much as well as too little is not optimum.

Hypothesis

The theme of this paper is that the relation between diversity and performance is likewise convex and complex and that too much diversity can be destabilizing as well as too little. Furthermore, my principal hypothesis is that optimum diversity is a function of the quality and quantity of energy flow. Low diversity may be optimum in ecosystems strongly subsidized by high quality auxillary energy flows and/or by large nutrient inputs, while a higher diversity may be optimum in ecosystems limited by the quality of energy input and/or dependent on internal nutrient recycling. Any positive correlation between diversity and stability is, therefore, a secondary rather than a primary relationship. Which is to say that quite stable systems in terms either of persistence in time or in terms of resistance to perturbation can have either a low or a high diversity, depending on the energy forcing function.

Methods

This tentative hypothesis is based on an analysis of about 150 censuses of major trophic and taxonomic components, such as vegetation, herbivores, aquatic bottom fauna, large taxonomic groups like arthropods, which play major roles in the ecosystems. Data on limited taxonomic groups such as lizards or butterflies were not included. Censuses were gathered from six ecological journals for the period of 1963–1973. In addition, the author's unpublished data on vegetation and arthropods in salt marshes, old fields, crops and forests were included. The total sample included a wide variety of natural, semi-natural, managed and cultivated ecosystems. A number of diversity indices were calculated, but for the purpose of this paper, only the reciprocal Simpson Index is used as follows: $1 - \sum(P_i)^2$ where P_i is probability for each species in terms of the ratio of its importance to the total of importance values. Numbers of individuals were generally the basis for importance, but where size of individuals varied widely within the component censused biomass or productivity was the basis for importance values. This form of the Simpson Index is scaled 0-1, where zero is lowest possible diversity (only one species) and highest diversities approach one.

Results

The frequency distribution of diversity indices for the ecosystems sampled was bimodal. Diversity was low (0.05 – 0.2) in one large group and moderately high (0.7 –0.85) in another large group. In no case were there large numbers of species relatively equal in importance, which would represent one kind of maximum. And there were relatively few cases of mid-level diversity (around 0.5). Ecosystems in the low diversity group included: (1) those stressed (i.e. degraded) by outside forces or inputs as in the case of polluted streams or bays, or those selectively managed so as to enhance dominance as in the case of croplands and commercial forests, and (2) those subsidized by large inputs of usable energy and/or materials as in the case of tidal marshes. The high diversity group included many natural, unsubsidized solar-powered ecosystems where sunlight, a low quality energy source in terms of per cent utility, is the only or chief energy source as in the case of many grasslands, upland forests and lakes in stable or nutrient-poor watersheds. High diversity also characterized communities in very stable physical environments as in the case of certain ocean bottoms and the wet tropics.

The pattern that seems to emerge is that the species matrix adapts to the strength and variety of energy input and the resource flows coupled with it. The strategy of nature is to diversify but not to the extent of reducing energetic efficiency. Furthermore, the quality of energy in terms of utility and low entropy is as important as the quantity. We can speculate that the optimum diversity is determined by both the kind and the level of energy input. When one or a few sources of high utility energy coupled with pumped-in growth-promoting substances are available in excess of maintenance needs, low diversity has advantages; a concentrated and specialized structure is more efficient in exploiting the bonanza than is a dispersed structure. It is perhaps under such conditions that high diversity is destabilizing (i.e. undesirable) as May's (1973) theoretical models show. High energy, low diversity systems can be quite stable both time-wise and in terms of resistance to perturbations if the input subsidies are regular or continue at the same level over long periods of time, as in the case of a tidal marsh. Under such conditions a low diversity of the order of 0.1–0.2 is optimum. But, of course, such systems will tend to 'boom and bust' when the subsidies fluctuate irregularly. Where energy is limiting or of low utility (as sunlight in the absence of other usable energy inputs) then a higher diversity of order of 0.7–0.8 appears to be optimum for the performance of the steady-state. Early developmental communities in secondary succession (but not necessarily in primary succession) tend to have low diversities which, according to our hypothesis, permits a better exploitation of accumulated and temporarily unused resources (see Odum 1969, for further discussion of diversity trends in succession).

Discussion

In the light of these results one must urge caution in the use of diversity as an index to pollution or other man-made disturbance. If the impacted system has a high diversity, then most stresses will certainly lower the diversity. But if the system has a low diversity to begin with, then a man-made perturbation may actually increase diversity ratios. In one of our studies (see Barrett, 1969) the evenness component of diversity of arthropods was actually increased after spraying with an insecticide, because the numbers of individuals of the strongly dominant species were reduced to a relatively greater extent than those of the rarer species.

Our general theory may be relevant to man's fuel-powered civilization. If we consider energy sources as very important 'species' then the developed countries of the world are now in the very low diversity category with 90 to 95% dependence on fossil fuel to run cities, industries and agriculture. Mankind has prospered in a material sense and his population has expanded rapidly as a result of his skill in exploiting one major source of energy. Abundant, high utility fuel such as oil is a positive feedback forcing function that has promoted a low diversity society whether the individual likes it or not. Now the problem is how to avoid the 'bust' as this major source declines. Should mankind strive for another dominant source, such as fusion atomic power, or should he conserve, diversify and work towards a powered-down steady-state? In view of the uncertain feasibility of another dominant high-utility source it would seem prudent for man to adopt the latter strategy. H. T. Odum (1973) argues that we must do so on the basis of increasing energy costs of converting declining fuels to useful work. We might envision the development of a distribution pattern of energy sources in the year 2000 something as follows: fossil fuel 50%, solar power 20%, atomic energy 10%, geothermal power 10%, water power 4%, and several other sources 6%. A Simpson diversity index for this mix comes to approximately 0.7, the level found in limited, steady-state natural systems. If we do not at least consider the diversity option and if fusion power cannot be harnessed within the next decade or so we may have no other choice but to suffer the 'bust' as predicted by the highly publicized '*Limits of Growth*' study (Meadows et al., 1972).

Summary

A study of the frequency distribution of diversity in a wide-ranging sample of ecosystems indicates the quantity and quality of energy input that determines the level of diversity within important functional groups such as trophic levels. Low diversity is characteristic of, and presumably optimum for, subsidized systems, while high diversity is selected for in resource-limited and steady-state systems. Therefore, a positive correlation between stability and diversity in not likely to be a cause-and-effect relationship. Caution is urged in the use of diversity indices for monitoring man's impact on natural systems since pollution or other stress may either increase or decrease diversity depending on systems energetics. The possible relevance of these theories to man's fuel-powered civilization is discussed.

References

Barrett, Gary W. P. 1969. The effects of an acute insecticide stress on a semi-enclosed grassland ecosystem. Ecology, 49: 1019–1035.

May, Robert M. 1973. Stability and Complexity in Model Ecosystems. Princeton Univ. Press.

Meadows, D. H. et al. 1972. The Limits of Growth. Universe Books, New York.

Odum, E. P., 1969. The Strategy of Ecosystem Development. Science, 164:262–270.

Odum, H. T. 1973. Energy, ecology and economics. AMBID (Royal Swedish Acad. Sci.), Vol. 2, No. 6.

Author's Address

Eugene P. Odum
Institute of Ecology
University of Georgia
Athens, Georgia 30602
U.S.A.

The concept of energy flow and nutrient flow between trophic levels

F. H. Rigler

Concept versus theory

The theme of this section is 'the flow of energy and matter between trophic levels'. This is an excellent theme with which to open our first congress because it gives us an opportunity to acknowledge one of the most important, and distressing, characteristics of ecology. Ecology is a discipline that seems to be more concerned with developing *concepts* than with producing *theories*. Evidence consistent with this statement is found in most ecological textbooks that are organized according to concepts. To drive the point home more effectively, it is proposed that the text of our plenary sessions is here published under the title 'Unifying Concepts in Ecology'. To see what is wrong with our concept-obsession, consider the dictionary meaning of the word. 'Concept' is usually defined as 'a general notion' or as 'an object conceived by the mind'. Fowler (1937) made a most appropriate comment when he wrote that 'concept' is 'a philosophical term, and should be left to the philosophers'. This is the point that I am trying to make. When ecologists use the word 'concept' they use it, not as a synonym for 'theory', but as a substitute for it. This is wrong because the job of science is not to collect a body of general notions. It is to produce theories. We can recognize a theory, as opposed to a concept, because the theory makes predictions about the external world, and predictions can be shown to be wrong. In fact, scientific theory *must* be potentially falsifiable by further observations. Concepts, being general notions, are more like the philosopher's premises, or the deductions from these premises, and, as such, are not predictive and are not falsifiable.[1]

The only point I want to make here is that most statements about trophic levels are, and can only be, statements about a concept. They are not theories because they cannot be falsified. However, this is not quite what one would

[1] Long ago Ehrlich & Holm (1962) published a criticism of biological concepts that is almost identical to mine although it may appear different. They objected to biologists equating concepts with facts, not with theory as I have done. However, their examples show that they really objected to the production of subsets of state variables that cannot be operationally defined. Obviously, general statements about undefinable sets of variables or objects cannot be falsified, and our objections to ecological concepts are consequently identical.

expect of an introduction to a session dedicated to the integration of previously isolated studies of energetics and materials in ecosystems. Therefore I will talk about ways and means of bringing the energy flow and nutrient cycle schools together, but I will do so, not necessarily because I believe this is an important goal, but because it provides a convenient vehicle with which to attack the concept in general, and the trophic level concept in particular.

The measurement of energy flow by conversion

We face two types of problem when we attempt to integrate our studies of energy flow and material flow in ecosystems. As an example of the first type, one could ask how we plan to measure energy flow. In aquatic systems, the only type in which I have studied energetics, we rarely measure energy flux. Instead, we measure the flux of some material such as CO_2, O_2, C or N, and then, by the judicious use of a standard conversion factor, we convert our material units into energy units. There is probably nothing wrong with this procedure provided we have accurate conversion factors and can measure material flow accurately. Unfortunately our conversion factors are usually suspect, and our measurements of material flux are not as reliable as we would like them to be. Radioactive isotopes facilitate the latter, but, even when the appropriate isotopes are available, the difficulties involved in making one accurate measurement are often pretty formidable. The possible sources of error in the ^{14}C method of measuring planktonic primary production would fill a page, and Conover & Francis (1973) have recently argued that the measurements of feeding and assimilation rates of small invertebrates are equally suspect. At a more fundamental level, I believe that the question raised by Edwards & Harris (1955), about the validity of using tracers to measure fluxes, still awaits a completely satisfactory answer. However, all of these problems are technical ones, and if it becomes important enough to us to solve them, we will do so.

The classification of ecosystem components with respect to cycling

The problem to which I want to call your attention now is much more difficult because different groups have studied the flow of energy and the flow of materials in the past, without considering the possibility of combining the two. Consequently they have not ensured that both used the same method of classifying the components of their ecosystem. If we are to combine the two we must ensure that the classification schemes are identical, or, at least, interconvertible. This will be much more difficult than improving our technology, because, to solve this one, some of us have to change our way of looking at nature.

Before talking about the origins and implications of the two classification systems, it might be worthwhile to define what I understand by a classification scheme, and explain why an accepted classification is important to us, and difficult to discard. Any natural system is too complex for us to predict the behaviour of every component. In fact, we could neither describe, nor recognize every component. Ecosystems are no exception. The potential array of component parts is too vast to be encompassed by our thoughts or theories. To make the system comprehensible, we simplify it by grouping its components into classes. However, the complexity of all systems, and the variety of the components are such that there are many equally valid methods of classifying the components. No one of these can be said to be correct. Yet the classification we adopt is all important. It usually reflects the way in which we were conditioned to see the world. In turn, it will influence the world view of those who use it. Past theory is subtly interwoven in the fabric of the system, as is the probability that it will inspire a new one. When Linnaeus classified fossils with rocks, rather than with animals or plants, he showed how strongly the theory of the permanence of species influenced his world view. When Francis Bacon set out to elaborate a theory of heat, he first classified all phenomena relevant to his interest. By including the heating of dung heaps in a list of, what we would now call, physical phenomena, he ensured that he would fail. In summary, classification is an essential, arbitrary process whereby we simplify nature, fossilize our current world view, and affect the probability of changing our theories. Because it embodies our world view, an accepted classification scheme is not easily discarded.

Now let us take a moment to imagine how we came to adopt two different schemes for classifying ecosystem components. Then we can examine both schemes in detail, and finally consider the nature of a possible common system.

The divergent schemes were probably adopted in the first place because ecosystems process energy and materials differently. Traditionally we say that energy flows unidirectionally, whereas materials cycle. This is only a crude approximation of the real difference, because energy and materials both cycle, and both flow out of ecosystems irreversibly. I define recycling as reutilization of inorganic substances or organic matter by an organism of the same trophic status as the organism that released the substances, or of a lower trophic status. Consequently, one needs only demonstrate that primary producers utilize dissolved organic compounds to prove that energy cycles. This has been done because not only the ability (Danforth, 1962), but also the need (Droop, 1962) of algae to utilize dissolved organic compounds is well documented. Aquatic animals can also utilize soluble organic compounds, protozoa utilizing them directly (Kidder & Dewey, 1951), and zooplankton making more effective use of them after they aggregate into particles (Gellis & Clarke, 1935; Baylor & Sutcliffe, 1963). However to know what fraction of this is truly recycled rather than transferred up the food chain we would also have to know the source of all the soluble organic compounds. Although my argument does not depend on it, it is interesting to note that whereas uptake of PO_4 by aquatic bacteria is accepted as part of the phosphorus cycle, the uptake of organic compounds by bacteria has not been treated as energy cycling. If we wished to make our terminology uniform, we could include bacterial uptake of organic compounds from the soluble pool as yet another example of energy cycling. Having shown that energy cycles in ecosystems, one needs only prove that materials do not cycle endlessly, but, like energy, are lost irreversibly from the system. This fact has also been adequately demonstrated and is well known. Terrestrial systems leach nutrients, and lakes lose them through the outflow and to the sediments. Lake Tahoe, which loses 93% of its incoming phosphorus to the sediments (Ludwig et al., 1964) gives an extreme example of loss to the sediments.

The efficiency of recycling

Although energy and materials both cycle they do so with very different efficiencies. (Efficiency here is defined as the ratio of the quantity recycled to the sum of the quantity recycled plus the quantity lost to the sink.) Energy cycles with a low efficiency whereas a substance, such as phosphorus, cycles with a high efficiency (Fig. 1). Although less dramatic than the traditional distinction, this way of looking at energy and material cycles has some interesting implications. It reminds us that an ecosystem will run down, not only if deprived of energy input, but also if the supply of any essential material is cut off. It directs our attention to the need for a continual input of nutrients as well as energy to maintain the system in a more or less steady state. By considering the form in which a nutrient and energy enter the system and the form in which they both recycle we can, perhaps, see the essential difference between the two types of cycle that led workers to adopt separate classification schemes.

Materials, such as phosphorus, can enter and cycle in the same form. Much of the supply comes in as orthophosphate ion and a major path in the cycle is

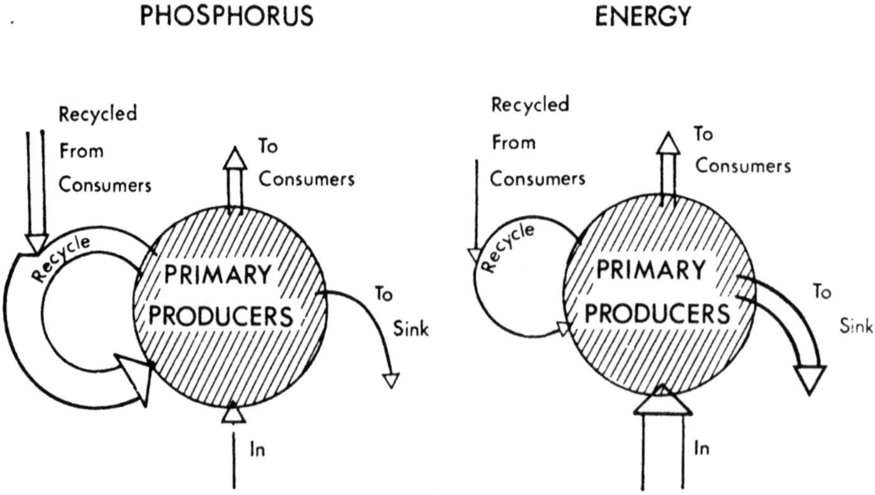

Figure 1. Diagrammatic representation of the difference between energy and nutrient cycles in an ecosystem.

ıly one obvious point of entry to the cycle. It is significant that the phosphate)ol is not only available to primary producers, but also to heterotrophic bacria and some animals. Energy does not enter and recycle in the same form.[1] enters as visible radiation and recycles only in the energy of chemical bonds. his gives us a choice of beginning our studies of the energy cycle either with .e incoming solar radiation, or with the pool of cycling organic matter. Because .e technology for measuring radiation and primary production was available ·st, studies of energetics began with the fixation of solar energy. However, the ıoice would probably have been the same even if labelled organic compounds ıd been available earlier because, unlike phosphorus, fixed energy does not pass ırough the soluble pool in a simple form. Every organic compound could be ; present in the pool, each one being formed and removed at a different rate om every other. Therefore the difficulty of measuring flux of energy through ıe soluble pool would have been much greater than that of measuring energy ‹ation. By beginning our work with energy fixation we tended to classify our 'stem components into primary producers and heterotrophs. The parallel ;tween this primary division and the pre-existing biological classification must ıve led to the conceptual subdivision of ecosystems into populations of bio-ıgical species.

To illustrate the subsequent divergence between the students of nutrients and ıergy I will compare the history of phosphorus studies in lakes with that of ıergetics studies.

he phosphorus cycle in aquatic systems

Work on the phosphorus cycle began before the need for continual input of hosphorus was recognized. The size of the phosphate pool was thought to show ıe supply of nutrient, and the earliest observational studies concerned annual uctuations in the size of this pool. Subsequently, experimental studies, of hich Einsele's (1941) fertilization of Schleinsee is an outstanding example, ıvolved alteration of the PO_4 pool. Einsele added massive quantities of superhosphate to the trophogenic zone and observed its disappearance (Fig. 2). here are two points to note. First, his method of subdividing the system was ;ry simple. The system comprised only three components, none of which was species or trophic level. These were PO_4 (the reservoir), colloidal P and parculate P. All organisms, as well as detritus, are included in one component. ;cond, Einsele did not sample the whole system. We can draw this conclusion ;cause only a small fraction of the added PO_4 appeared in suspended particles ter the first addition, and although there was a quantitative movement of the

This applies only to the common types of system in which primary production/ ıtochthonous input ≫1. In systems such as deep caves the cycles of energy and materis are more similar.

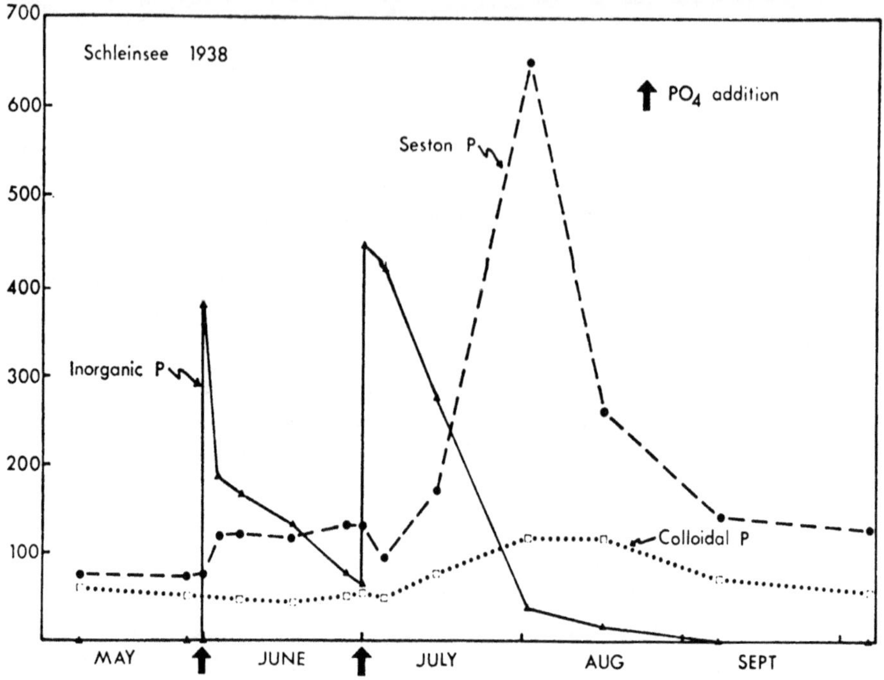

Figure 2. Changes in size of different phosphorus compartments in Schleinsee following additions of PO_4–P to the surface waters. Concentrations expressed in μgl/P. Data were calculated from Einsele (1941).

second dose of PO_4 into particles, the total quantity of P in the trophogenic zone almost returned to normal within 3 months. The results obviously require more, or different, components in the classification scheme. Shortly after this work, radioactive P became available and two groups (Hutchinson & Bowen, 1947; Coffin et al., 1949) quickly took advantage of the opportunity to study P cycling without disturbing the system.

Hutchinson & Bowen (1950) showed that ^{32}P was equilibrated between solution and particulate matter (seston) within a week. They also showed that P moves from the open water to littoral vegetation, and from the trophogenic zone to the tropholytic zone. After their experiment the lake was pictured as in Figure 3. Notice that their classification scheme is not only adequate to explain their own results, but also provides the component missing from Einsele's scheme. Next, Rigler (1956) showed that there is a rapid exchange between seston and PO_4 and postulated a similar exchange between the pelagial and littoral portions of the trophogenic zone. Finally Chamberlain (1968) showed that the kinetic data did not require that a return of ^{32}P from the littoral be postulated. He also showed that loss of ^{32}P from the pelagial, trophogenic zone was not the expected, simple, exponential rate. In the first week it was higher than in the following weeks. This observation forced him to modify the model by subdividing one component. Because he also showed that particles bigger than 70 μ take at

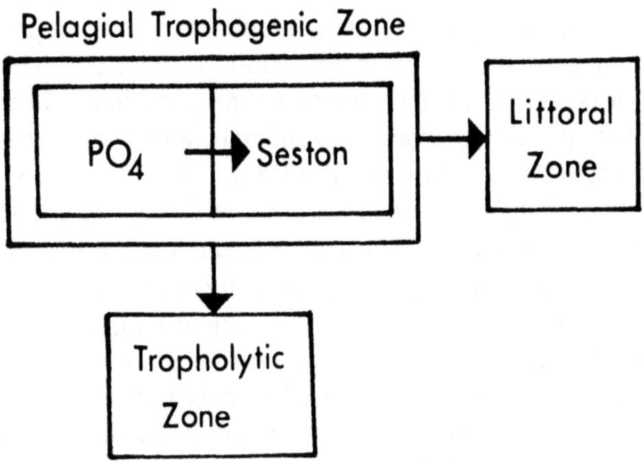

Figure 3. Phosphorus compartments in a lake as visualized after the work of Hutchinson & Bowen (1950).

least a week to reach isotopic equilibrium he chose to subdivide it as in Figure 4. Note that one compartment was added and one flux deleted from littoral to pelagial to make the model fit the new data.

Subsequent observations forced us to complicate further the classification scheme (Rigler, 1973), but I will not describe these in detail because the recent developments merely reinforce the three points I have tried to make. The first, is that the scheme for classifying components of the system was flexible. Second, the components are operationally defined. The particles caught on a 70 μ net do not coincide with any ecological concepts, but statements about this compartment can be tested. Third, the biological species and the trophic level do not appear in any of the classification schemes used to describe nutrient cycling.

The energy cycle and the trophic level concept

The direction taken by students of ecosystem energetics was very different, Their starting point was primary production, possibly, as I suggested earlier.

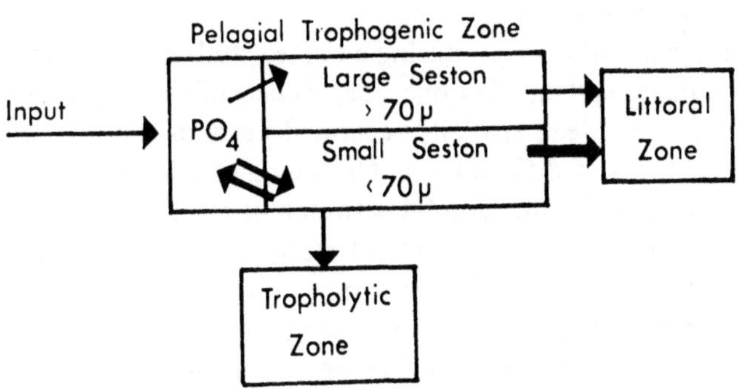

Figure 4. Phosphorus compartments in a lake as visualized after the work of Chamberlain (1968).

because this process transforms energy into a form in which it can cycle, or, possibly, because primary production is the most easily quantified process. Having adopted a biological taxon as their first compartment the energy cyclists looked next to the consumption of primary producers by animals, and immediately immersed themselves in what has frequently been called the 'bewildering complexity' of ecosystems. The system became complex because the flux of energy from primary producers to the animals that consume them was not directly measurable. To determine this flux, the grazing rate of each animal species was measured, and the individual rates were summed. This method imposed on us a classification scheme in which each species of animal was a component. However, studies such as Hardy's (1924) work on the feeding habits of the herring must have convinced many ecologists that this classification system comprised an unwieldy number of components, and attempts were soon made to simplify the species-based scheme. Juday (1940) tried to simplify it by subdividing his lake into phytoplankton, zooplankton, bottom flora and fauna, and fish, but his classification system did not have the general applicability or conceptual consistency of Lindeman's (1942), trophic level scheme that has been generally adopted. In this scheme all organisms in the designated ecosystem are grouped, regardless of size or location, according to their trophic relation to the primary energy source.

The appeal of this trophic level concept was immediate and has strengthened for 32 years despite the fact that the scheme was known to be inadequate from the beginning. Who would have dared to assign Hardy's herring to any particular trophic level? Lindeman himself stated that the farther removed an animal is from the initial source of energy, the smaller is its chance of deriving nutrition from one trophic level. Ivlev (1945), Odum (1959) and Mann (1969), all strong supporters of the trophic level scheme, admitted this flaw. Others who recognized it were more dissatisfied and tried to modify the use (Cummins et al., 1966) or structure (Wiegert and Owen, 1971) of the concept, and Darnell (1961) unsuccessfully attempted to replace it with a trophic spectrum scheme. Nevertheless a glance through this book shows that the classification by trophic levels has thrived on exposure of its inadequacy.

Let us now return to the question of producing a unified study of phosphorus and energy cycling in aquatic systems. On one side, models of the phosphorus cycle have described fluxes between spatially separated parts of the lake and between solution and particles of different size. On the other side, energy flux is measured from one species to another, and description of results is simplified by grouping species into trophic levels. At present there is no correspondence between the two classification systems. This situation must be corrected if we are to combine the two types of study. I do not know how we will do this. In fact, until it has been accomplished successfully, no one can know. I shall merely suggest that the obvious method, in which the minority, who work on nutrients, should reject their scheme, and come into line with the majority, who work on energy cycling, is not the solution. This is because the weakness of the trophic level scheme, to which I have already alluded, is not trivial. In fact, it is symptomatic of the sickness of modern ecology.

A revealing argument about the trophic level concept

In the time remaining, I will explain this statement by examining one example of the trophic level concept in action. This concerns the minor controversy that arose from the work of Hairston, Smith & Slobodkin (1960). They wrote an interesting and valuable paper in which they attempted to rise above the maze of detail that entrances most ecologists, and make general statements about the factors that set an upper limit to the biomass of large groups of organisms. Their groups were the trophic levels; producers, herbivores, carnivores and decomposers, and their argument required every organism to fit into one of these categories.[1] This is the only flaw with which I shall be concerned today.

This flaw was recognized and attacked long ago by Murdoch (1966) in what will, hopefully, be recognized one day as one of the most important ecological papers ever written.[2] The value of Murdoch's work is that it is an attempt to make us consciously aware of the criterion by which we should judge ecological statements. This criterion is very simple. Scientific theories can never be shown to be true; they can only be shown to be false. To be potentially falsifiable they must make predictions about the future state of some system, its state being the description of all of its state variables or observable properties. Furthermore they cannot permit or predict all possible states of a system because, if they do, their predictions cannot be falsified. A statement of this type could be called a concept, but not a theory or hypothesis. A statement that prohibits only one of all the conceivable states is a theory (although not a particularly useful one) because it would be falsified by the occurrence of the one prohibited state. A theory that prohibits all but one state is the most useful. It is also the most easily falsified. Ecological statements, if they are to be considered scientific, must be susceptible to falsification.

[1] They actually vacillate between statements about trophic levels and statements about populations. This uncertainty led Ehrlich & Birch (1967) astray, and made their objections to Hairston, Smith and Slobodkin's hypothesis susceptible to refutation because Ehrlich and Birch attempted to falsify the hypothesis with observations concerning populations. In their reply to Ehrlich and Birch, Slobodkin, Smith & Hairston (1967) remove the ambiguity by stating that they were not, 'making statements about most herbivores or most carnivores, but about these trophic levels as wholes'.

[2] It is distressing that the importance of Murdoch's work has not yet been generally recognized. Whereas Citations Index lists 28 citations to Hairston, Smith & Slobodkin (1960) in the first 7 years after publication, it lists only 13 to Murdoch in the comparable period. Furthermore, excluding Ehrlich & Birch (1967) and Slobodkin, Smith & Hairston (1967), none of the authors citing Murdoch was impressed by the crucial issue raised by Murdoch. They all made nothing more than a casual reference to his criticism.

The weakness that Murdoch spotted in Hairston et al. (1960) was that it makes a number of general statements that appear to be theories but are not falsifiable. This is because they refer to trophic levels as if they were state variables of the ecosystem, and the trophic level is a concept that cannot be related to any quantifiable property of natural systems. The reason why the trophic level is not a state variable is that, as others recognized and I mentioned earlier, many species cannot be assigned to a single trophic level. Slobodkin et al. (1967) were not unaware of this, but dealt with it by saying that 'the widespread occurrence of organisms that are not clearly assignable to [a] trophic level does not falsify the conclusions reached'. They may be correct, but they have answered the wrong objection. The fact that species do not fit into the categories makes it impossible to falsify the conclusions.

Unfortunately the matter cannot be left here because Slobodkin et al. (1967) disagreed with Murdoch, and argued that every one of Hairston et al.'s (1960) conclusions was falsifiable. I believe they failed in every case and furthermore that failure was inevitable except, perhaps, in the case of producers. I tentatively exclude producers because, for most terrestrial systems, we could almost certainly agree on the species to include in this category. If so, biomass of producers is a concept that could be made to coincide with an observable property of ecosystems. It is much more difficult to imagine how this could be true of any of the other trophic levels. To make the difficulty more obvious, the hypotheses concerning herbivores (Slobodkin et al., conclusions 3 and 4) can be reworded to state that the average biomass of the herbivore trophic level is negatively correlated with the rate of biomass loss due to predation plus parasitism, and is independent of total annual net primary productivity. I merely added 'average biomass' because it can be quantified as ash-free dry weight or by any one of a number of reproducible measurements, and used net primary productivity as a measure or rate of supply of the resource. One could contemplate several tests of this hypothesis: the immense difficulty of performing any experiment should be of no concern, provided it is theoretically possible. First, one could stimulate or inhibit net primary productivity, and show a positive correlation with herbivore biomass. Second, one could demonstrate the existence of an environmental variable that changes the biomass of herbivores without affecting net primary productivity or predation plus parasitism. Alternatively, one could experimentally alter total predation plus parasitism and show that within some predetermined limits, herbivore biomass is unaffected. Assuming, for now, that we could agree on the temporal and spatial constraints that would have to be added to our hypothesis, and neglecting the fact that some of these experiments are subject to more objections than others, they can all be disqualified by one common fault. In every one the average biomass of the herbivore trophic level must be measured. This is exactly what everyone, including Slobodkin et al. (1967), agrees we cannot do. Too many animal species fall in a spectrum between herbivores and carnivores; others fall between decomposers and herbivores;

still others combine characteristics of all three levels. Unless we decide to agree on a workable, operational definition of 'herbivore', the biomass of herbivores will remain a concept, and statements about this concept will be untestable. The remaining conclusions of Hairston et al. (1960) concerning predators and decomposers are subject to the same objection and, provided you have accepted the argument against the use of the herbivore level, statements about the other two can also be rejected as concepts, not hypotheses.

Conclusion

What I have shown from statements about limitations of the biomass of trophic levels applies equally to statements about energy flow between trophic levels. These are also unfalsifiable. This is why I do not think it would be profitable for the study of material cycles to go the way of the study of energy cycles.

In closing, I want to ask one last question. Why do we care about bringing the two types of study together? It must be that the two lines, taken individually, have not spawned useful theories. We now have the feeling that if we combine them, something useful is bound to appear. I want to suggest that we might be better advised to concern ourselves with making our statements about nutrients and energy falsifiable than with combining them. If we are not careful, the combination may be productive of nothing but more concepts.

References

Baylor, E. R. and W. H. Sutcliffe Jr. 1963. Dissolved organic matter in seawater as a source of particulate food. Limnol. Oceanogr. 8: 369–371.

Chamberlain, W. M. 1968. A preliminary investigation of the nature and importance of soluble organic phosphorus in the phosphorus cycle of lakes. Ph.D. Thesis, Univ. of Toronto, 232p.

Coffin, C. C., F. R. Hayes, L. H. Jodrey and S. G. Whiteway. 1949. Exchange of materials in a lake as studied by the addition of radioactive phosphorus. Can. J. Res. 27: 207–222.

Conover, R. J. and V. Francis. 1973. The use of radioactive isotopes to measure the transfer of materials in aquatic food chains. Mar. Biol. 18: 272–283.

Cummins, K. W., W. P. Coffman and P. A. Roff. 1966. Trophic relations in a small woodland stream. Verh. int. Ver. Limnol. 16: 627–638.

Danforth, W. F. 1962. Substrate assimilation and heterotrophy. In: Physiology and biochemistry of algae. R. A. Lewin ed. New York, London, 99–123.

Darnell, R. M. 1961. Trophic spectrum of an estuarine community based on studies of Lake Pontchartrain, Louisiana. Ecology 42: 553–568.

Droop, M. R. 1962. Organic micronutrients. In: Physiology and biochemistry of algae. R. A. Lewin ed. New York, London, 141–159.

Edwards, C. and E. J. Harris. 1955. Do tracers measure fluxes? Nature, Lond. 175: 262.

Ehrlich, P. R. and L. C. Birch. 1967. The 'balance of nature' and 'population control'. Amer. Natur. 101: 97–107.

Ehrlich, P. R. and R. W. Holm. 1962. Patterns and populations. Science, 137: 652–657.

Einsele, W. 1941. Die Umsetzung von zugeführtem, anorganischem Phosphat im eutrophen See und ihre Rückwirkungen auf seinen Gesamthaushalt. Z. Fisch. 39: 407–488.

Fowler, H. W. 1937. A dictionary of modern English usage. Oxford. 742p.
Gellis, S. S. and G. L. Clarke. 1935. Organic matter in dissolved and in colloidal form as food for *Daphnia magna*. Physiol. Zool. 8: 127–137.
Hairston, N. G., F. E. Smith and L. B. Slobodkin. 1960. Community structure, population control, and competition. Amer. Natur. 94: 421–424.
Hardy, A. C. 1924. The herring in relation to its animate environment. I. The food and feeding habits of the herring with special reference to the east coast of England. Min. Agr. Fish. Fisheries Investigation Series II 7(3): 1–53.
Hutchinson, G. E. and V. T. Bowen. 1947. A direct demonstration of the phosphorus cycle in a small lake. Proc. Nat. Acad. Sci. 33: 148–153.
Hutchinson, G. E. and V. T. Bowen. 1950. Limnological studies in Connecticut. IX. A quantitative radiochemical study of the phosphorus cycle in Linsley Pond. Ecology 31: 194–203.
Ivlev, V. S. 1945. Biologicheskaya produktivnost' vodoemov. Uspekhi sovr. biologii 19: 98–120.
Juday, C. 1940. The annual energy budget of an inland lake. Ecology 21: 438–450.
Kidder, G. W. and V. C. Dewey. 1951. The biochemistry of ciliates in pure culture. In: Biochemistry and physiology of Protozoa, A. Lwoff ed. Academic Press, New York, 323–400.
Lindeman, R. L. 1942. The trophic-dynamic aspect of ecology. Ecology 23: 399–418.
Ludwig, H. F., E. Kazmierczak and R. C. Carter. 1964. Waste disposal and the future of Lake Tahoe. J. Sanit. Eng. Div. ASCE 90, SA 3. Proc. Paper 3947: 27–51.
Mann, K. H. 1969. The dynamics of aquatic ecosystems. Advances in Ecological Research 6: 1–81. Academic Press, New York.
Murdoch, W. W. 1966. 'Community structure, population control, and competition'—a critique. Amer. Natur. 100: 219–226.
Odum, E. P. 1959. Fundamentals of ecology. Philadelphia, W. B. Saunders Co. 546p.
Rigler, F. H. 1956. A tracer study of the phosphorus cycle in lakes. Ecology 37: 550–562.
Rigler, F. H. 1973. A dynamic view of the phosphorus cycle in lakes. In: Environmental phosphorus handbook,E. J. Griffiths et al. ed. Wiley, New York, 539–572.
Slobodkin, L. B., F. E. Smith and N. G. Hairston. 1967. Regulation in terrestrial ecosystems, and the implied balance of nature. Amer. Natur. 101: 109–124.
Wiegert, R. G. and D. F. Owen. 1971. Trophic structure, available resources and population density in terrestrial vs. aquatic ecosystems. J. Theor. Biol. 30: 69–81.

Author's address:

F. H. Rigler
Department of Zoology
University of Toronto
Toronto
Canada

Principles of energy and material exchange in ecosystems[1]

D. E. Reichle, R. V. O'Neill & W. F. Harris

Introduction

Ecological systems are conceived as energy-processing units, ordinarily not radiant energy-limited, but regulated by the availability of essential nutrient elements and water, and constrained by climate. Energy is the fuel through which ecological processes operate, but the rates at which processes occur are controlled in natural systems by nutrient availability. It is hypothesized that ecosystems function to expend readily available energy to minimize the constraints imposed by limiting nutrients and water. Ecosystems display adaptive control mechanisms associated with heterotrophic activities (e.g. grazing and decomposition) that lead to conservation of nutrient capital. Therefore, patterns of *energy flow* and *element cycling* in ecological systems cannot be interpreted independently without the chance of introducing erroneous conclusions about ecosystem function.

A foundation for discussing the principles of ecosystem function must begin at a very basic level with the definition of a system. Numerous definitions have been offered in the context of biological systems (Milsum, 1966; Watt, 1966; White, 1969; Smith, 1970; Patten, 1971). Weiss (1971) distinguishes a 'unit' as a composite which retains some identity over time. A system is specified as a complex unit in which the components interact to preserve the system and restore it following non-destructive disturbances. Weiss specifically notes that a system, and particularly a biological system, will represent some level in a hierarchy. The system will be composed of subsystems and will itself be a subsystem of some higher level of organization. Therefore, an analytic approach to understanding any system at a given level of organization should compare the system's properties with those of other systems in the same hierarchy. We will attempt to adopt this hierarchical epistemology in our discussion of the ecosystem.

[1] Research supported in part by the U.S. Atomic Energy Commission under contract with the Union Carbide Corporation, and in part by the Eastern Deciduous Forest Biome, US-IBP, funded by the National Science Foundation under Interagency Agreement AG-199, BMS69-01147 A09 with the Atomic Energy Commission, Oak Ridge National Laboratory.
Contribution No. 160 from the Eastern Deciduous Forest Biome, US-IBP. Publication No. 667, Environmental Sciences Division, ORNL.

Fundamental to exploring the functional coupling of energy and elements in ecological systems is the principle that energy *flows* through ecosystems, while elements can *cycle* within the system (albeit various degrees of nutrient leakage do occur). Consequently, it is sometimes thought that these two concepts of flows and cycles are mutually incompatible and that energy flow cannot be related to element cycling. We do not believe that this is so. The problem arises in the conceptualization of flows of energy and cycles of elements in ecosystems. The former are traditionally portioned for analysis according to trophic levels which are not directly compatible with the functional ecosystem processes responsible for element cycling. Difficulties in applying trophic level concepts to ecosystem functions lie partially in relating energetic functions of populations to the processes of ecosystems. (An example is the consumer trophic level which includes both herbivores and decomposers—each with unique ecosystem roles in element cycling).

The change in our conceptual thinking needs to take us to evaluating the energetic 'costs' of various element cycling processes in ecosystems. In this concept, flows of energy and cycles of elements are compatible. Energy is a relatively unlimited resource which the ecosystem expends to conserve limiting nutrient elements. We know of no ecosystems that are entirely radiant energy-limited. In this context, it is often impossible to deduce 'strategies' of ecosystems solely from either a nutrient or an energetic perspective, since each may be counter-intuitive—e.g. an ecosystem may be very inefficient in energy utilization in order to be very efficient in nutrient utilization.

The ecosystem concept

Definition of the system

A system is a complex of interacting subsystems which persists through time due to the interaction of its components. The system possesses a definable organization, temporal continuity, and functional properties which can be viewed as distinctive to the system rather than its components. In what sense then may we designate an ecosystem as being a system? Does it indeed possess these attributes? Ecosystems retain identity in geographic perspective, even though dramatic shifts may occur in climatological and geological variables. The ecosystem is continuous in time and maintains this continuity in spite of disturbances. The entire complex of phenomena designated as secondary succession (Odum, 1969) shows a capability of the ecosystem to maintain its identity. The ecosystem possesses a definite organization in its trophic structure and, indeed, this trophic structure remains relatively constant in spite of disturbances (Heatwole & Levins, 1972) and in spite of vastly different geographic settings including numerous examples of convergent evolution. Numerous functional properties can be seen as distinctive attributes of the ecosystem (Reichle, 1975). It has long been established that there is firm ground for dealing with the ecosystem as a distinct object of study and as a system with distinct functional properties.

Hierarchy of biological systems

Since the ecosystem qualifies under this definition of a system as offered by Weiss (1971), we will proceed with the hierarchical methodology. This involves examination of the various levels of organization in the biological hierarchy, of which the ecosystem is a part, in order to identify some aspects of system organization which are common to all levels of organization in that hierarchy. Thus, without overextending the analogy, we will explore some fundamental properties of biological systems.

Perhaps the most fundamental properties which can be identified with respect to biological systems are the tendencies to persist and the capability to grow in spite of fluctuation in the environment. The tendency to persist and grow can be identified at each level of organization, and frequently can be identified as superseding the persistence of component subsystems. At each level of organization, the system persists and grows even if component subsystems must be sacrificed. The cell persists even though specific chemical constituents are replaced. The organism persists and grows even though individual cells perish. The population expands for periods of time far in excess of individual life spans. The ecosystem persists even though individual populations are replaced. Although the mechanisms underlying the various levels of organization may differ, persistence and growth appear as fundamental properties of living systems.

Growth toward maximum persistent biomass

Further elaboration can illustrate the ways in which the properties of persistence and growth manifest themselves at the ecosystem level. The range of environmental conditions in an area will exclude individual populations of plants and animals. Populations which are capable of survival and reproduction will form the elements from which the system is established. The populations will form interactions and establish feedbacks such that the ecosystem persists and grows. Since environment is not constant, fluctuations will occur over daily, seasonal, annual or longer time periods. System interactions will tend to counteract these fluctuations. The system will grow in a constant environment to a maximum *potential* biomass at equilibrium with the environment (Fig. 1). This growth process can be interrupted by disturbances or fluctuations in the environment so that the system oscillates at some lower level around the maximum *persistent* biomass. Figure 2 illustrates how the average biomass of the system will be reduced as the severity of the environment, i.e. as mean climatic factors, deviate from some optimal level. Superimposed on the severity of the environment will be the severity and frequency of disturbances or fluctuations around these mean environmental conditions. As severity or frequency of the disturbance increases, there will be a reduction in the maximum biomass. This is defined as the maximum persistent biomass.

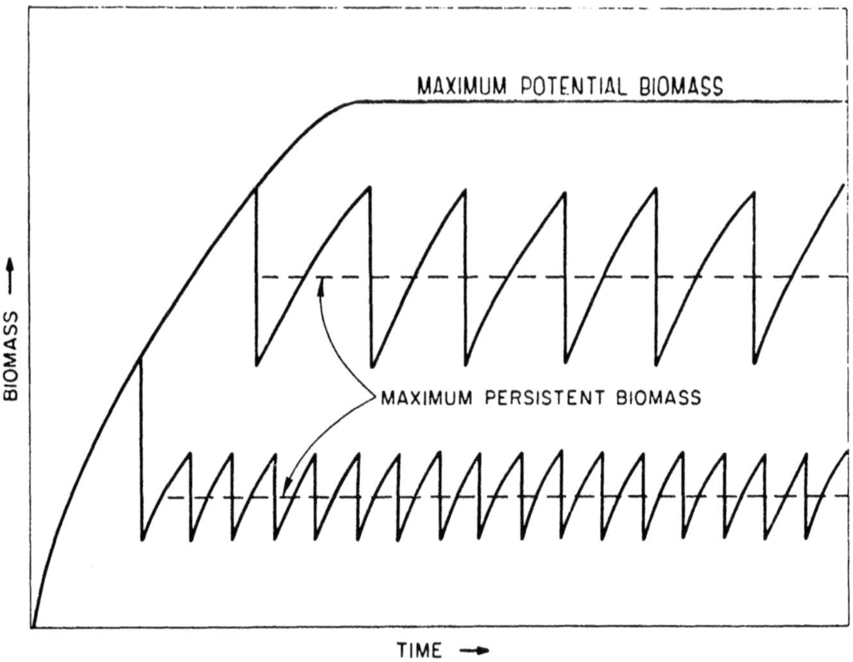

Figure 1. Changes in ecosystem standing crop through time. In the absence of environmental fluctuations, the system will tend to grow to some maximum potential biomass determined by the severity of the habitat. The effects of perturbations and subsequent recovery will cause the system to fluctuate about a lower level, the maximum persistent biomass. The more frequent and the more severe the perturbations, the lower will be the persistent biomass.

Without maintaining that these functional properties are the only attributes worth studying in the system, we have tried to show that homeostatic mechanisms which lead to maximum persistent biomass are related to very fundamental properties of the ecosystem level of organization. We suggest that ecosystem properties can be properly studied without postulating new and ethereal mechanisms of evolution and without endowing the ecosystem with unjustifiable attributes. Therefore, we can on this basis proceed with a description of hypothesized mechanisms operating at the ecosystem level, which govern energy flows and nutrient cycles so as to permit persistence and growth of these systems.

Analysis of ecosystems

The general properties of persistence and growth common to all biological systems can be seen on the ecosystem scale to constitute growth toward the maximum amount of living tissue consistent with the average environment and the fluctuations around this average. Homeostatic mechanisms at the ecosystem level of organization are unique in permitting persistence and growth of the total system even though individual populations do not survive. These mechanisms, taking the form of interactions between component subsystems, form the proper objects of study for ecosystem analysis and define a unique set of problems not addressed at other levels of biological investigation.

Historically, ecological theory has focused on the individual organism and the population. We believe that it is possible to investigate the interactions between populations in the ecosystem as these relate to the persistence of the total system. The ecosystem can be addressed as a unit without contradicting the principle that individual populations react to environmental gradients in a relatively independent manner. Natural selection on a population can occur from other populations in the system: e.g. competition, development of defensive mechanisms by plants and prey organisms, or evolution of flower parts in plants dependent upon animals for pollination. These and countless other examples illustrate that the population evolves within the context of its biological environment. Unless populations in the ecosystem contribute to specific vital system functions such as photosynthesis or cycling of nutrients, strong negative selective

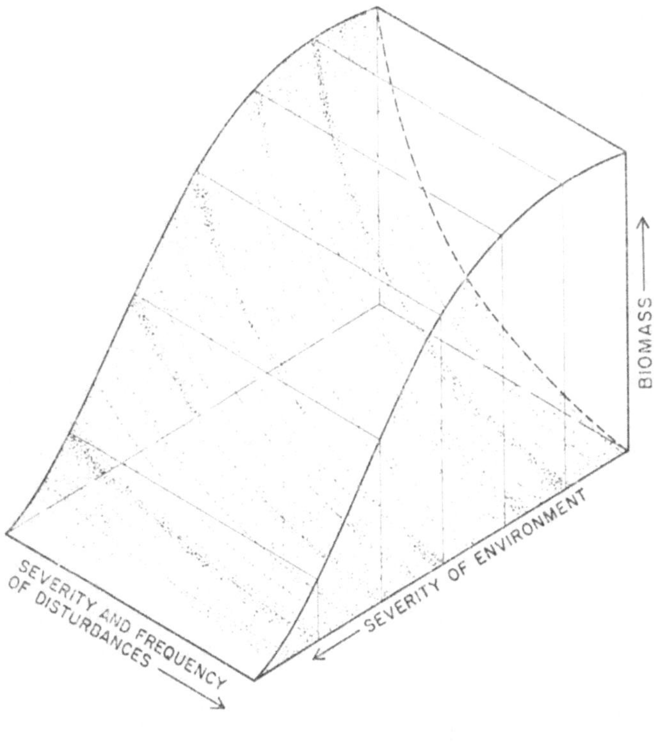

Figure 2. Mean ecosystem standing crop as a function of physical environment. The more severe the environment the lower the maximum potential biomass. When severity of perturbations is superimposed a still lower maximum persistent biomass results. The exact shape of the response surface is arbitrary, although the monotonic decrease is consistent with observation.

pressure is exerted upon the individual populations by the system. Whatever individual population responses are established, the populations establish homeostatic feedback mechanisms so that the ecosystem survives and grows to a maximum persistent biomass. In this inquiry into the functional properties of ecosystems, we discuss homeostatic mechanisms common to this level of organization—emphasizing basic principles rather than causal relationships.

Principles of ecosystem function

If factors causing persistence of ecological systems in a fluctuating environment serve as natural selection pressures on component populations, we should be able to identify characteristics at the ecosystem level of organization which are based on component population interactions, but which contribute to the persistence of the ecosystem of which they are a part. The specific nature of the mechanisms underlying ecosystem function will vary; the taxonomic diversity of biota and the continuum of climatic extremes and variations support this intuition. What we ultimately seek to understand is the causality of evolutionary convergence to yield maximum persistent biomass as a basic ecosystem property. Thus, it is not our intent to reintroduce concepts, processes and properties unique to the ecosystem level of organization. Our objective is to identify specific ecosystem attributes which relate to a single ecosystem phenomenon—persistence.

The energy base

The ultimate source of energy which drives the ecosystems in our biosphere is the sun. The availability of fixed solar energy is implicit in the definition of an ecosystem (Odum, 1971). There may be ecosystems (e.g. woodland streams) with a minimal authochthonous energy base. These ecosystems are not exceptions, since they require subsidy of chemical energy from outside the physical boundaries of the ecosystem. Thus, the fixation of solar energy is essential to growth and maintenance of ecosystems. Autotrophic populations capable of survival in a particular climatic regime provide the energy base and display many evolutionary adaptations to optimize use of water and light in providing a firm energy base for support of secondary trophic structure.

Homeostatic mechanisms on the ecosystem level of system organization relate to collective population adaptations of autotrophs which ensure the continual existence of a secure energy base. Two basic adaptations are apparent: 'small-fast' and 'big-slow' functional components. Many plant populations may be present, each with characteristic ecological amplitude, but each with capability of rapid growth and reproduction under favorable conditions. Many aquatic ecosystems which rely on planktonic populations display this adaptation predominantly. The second adaptation relies on energy fixation by individuals which attain great bulk and elaborate morphological and population structure and exhibit relatively slower reproduction. Forests exemplify predominance of this adaptation.

Besides adopting (predominantly) one of these adaptations, many ecosystems also have alternative capabilities to maintain a firm energy base. Thus, forest ecosystems include an herbaceous layer capable of rapid growth and reproduction (Taylor, 1974). Forests also contain populations of saplings in the subcanopy strata capable of rapid growth to canopy stature in the event that a canopy opening occurs. Lake ecosystems with a dominance of phytoplankton production may also contain a zone of rooted or floating macrophytes. Thus, the persistence of a firm energy base in a fluctuating environment relies on the opportunism afforded by the mix relative to that ecosystem of 'fast' and 'slow' autotrophic populations.

A reservoir energy base

Despite the flexibility in maintaining a carbon base afforded by a mix of autotrophic adaptations, the probability exists that environmental extremes (biotic or abiotic) may perturb the energy base of an ecosystem. Intuitively, if ecosystems are persistent, some ecosystem alternative must exist which can provide a reservoir energy base. This energy base may be of suboptimal quality, could be difficult to mobilize, and might serve multiple functions besides acting as an energy base. In an earlier paper (O'Neill et al., in press), we have proposed that this energy reservoir should have at least two characteristics. It likely would be large (perhaps compensating for its suboptimal quality) and it would exhibit slow response times such that short-term fluctuations in environmental conditions would have minimal effect.

Many ecosystems representative of a wide range of structural and climatic conditions do indeed possess a component with the characteristics of large size and slow response (Table 1). It is of interest to note that this energy reservoir

Table 1. Slowest component and turnover times for a variety of ecosystems (after O'Neill et al., in press).

	Slowest component	Turnover time (Years)
Tundra		
Whitfield 1972	Soil organic	340.6
Deciduous Forests		
Duvigneaud & Denaeyer-DeSmet 1970 (calcium)	Soil	108
Olson 1965 (radiocesium)	Soil	25
Satchell 1971 (energy)	Soil organic	76
Harris et al. 1975 (carbon)	Structural organic matter	155
Henderson and Harris 1975 (nitrogen)	Soil organic	109
Tropical Rain Forests		
McGinnis et al., 1969 (elements)	Soil	41
Odum 1970 (energy)	Soil	26
Peat Bog		
Gore and Olson 1967 (biomass)	Anaerobic peat	526

may be composed of two structural components. In a temperate deciduous forest woody supportive structure and soil organic matter (excluding detritus of recent origin and roots) both have comparably long residence times (see for example Harris et al., 1975). These several sources comprising an energy reservoir suggest other properties related to such a resource. In the face of extreme environmental fluctuation, this alternative energy base represents a source of energy to a remaining or invading trophic level (saprovores) whose activities release essential nutrient elements necessary for ecesis of new autotroph populations. The concept of an energy reservoir implies energy storage as an investment against unlikely catastrophic events. Energy storage represents the energetic costs of maintaining essential elements in a form and proximity which can be reached, remobilized and recycled by the ecosystem. In this sense, the 'costs' of such a reservoir are those of long-term system maintenance of essential element resources, and support the contention that ecosystems are energy processing units regulated by, and inextricably connected to, the cycles of nutrient elements in the system.

Element recycling

An array of elements are required to support growth and maintenance of ecosystems, and unlike energy, these chemicals can be recycled. Many mechanisms at species and population levels can be identified that conserve elements for short-term recycling. These mechanisms have associated energy costs. Examples such as restitution-retention (Duvigneaud & Denaeyer-De Smet, 1970; Henderson & Harris, 1975; Wells & Jorgensen, 1975), evergreenness (Monk, 1966) and mycorrhizal associations (Harley, 1959) are illustrative of these mechanisms. If the ecosystem level of organization transcends the temporal bounds of component populations, then ecosystem function should encompass mechanisms that enhance conservation of essential elements which otherwise would be unavailable to the ecosystem.

The hypothesis has been posed that given adequate water and energy resources, the level of maximum persistent biomass in an ecosystem will be determined by the supply of essential elements (O'Neill et al., in press). Where energy or water are limiting, the recycling potential may be reduced, since a biomass accumulation sufficiently large to exhaust the element resources cannot be maintained under these conditions. Therefore, in terrestrial ecosystems, element-conserving mechanisms should be best developed in warm, mesic forest ecosystems where energy and water are abundant.

Nutrient storage is another key role of the energy base in ecosystem persistence. The attributes of an alternate energy base are fundamental for effective nutrient recycling. Large organic mass provides capability to store quantities of nutrient elements; slow turnover of the alternate energy base maximizes the probability that elements will be retained within the system. Build-up of an organic matter pool requires a net primary production exceeding the energy demands of living tissues and the energy 'costs' of mineralization by decomposers of this pool. The cost of such storage is equal to the caloric equivalent of the organic matter accumulation (net ecosystem production). Thus, the energy content of the maximum persistent organic matter is a part of the long-term maintenance, or persistence, of an ecosystem. Accumulation of inactive organic matter represents an energy cost insuring availability of essential elements.

Element storage alone does not provide for ecosystem persistence. The missing attribute is a means of regulated remobilization of this nutrient resource. This mechanism is provided by decomposers. Catabolic activity of decomposers releases element reserves in forms and at rates available to autotrophs. Thus, this aspect of ecosystem function embodies not only interactions of populations of the same trophic level, but also interactions among trophic levels. To illustrate the effectiveness of this recycling operation, we shall reference specific research on ecosystem carbon cycles (Reichle et al., 1973a; Harris et al., 1975) and nitrogen nutrition (Henderson & Harris, 1975) summarized in Table 2. Although other chemical species are conserved and cycled in the system, nitrogen is of particular interest since it exists naturally in gaseous form and ordinarily must

Table 2. Comparison of turnover times for carbon, nitrogen and calcium in temperate deciduous forests (Tennessee) (from O'Neill et al., in press).

	Turnover time (Yr)		
Component	Carbon[a]	Nitrogen[b]	Calcium[c]
Soil	107	109	32[e]
Forest Biomass[d]	155	88	8
Litter (01 + 02)	1.12	<5	<5
Total	54	1815	445
Decomposers	0.01	0.02	?

[a] Data based on carbon metabolism of yellow poplar forest (Harris et al., 1975 Reichle et al., 1973a).
[b] Data based on nitrogen budget for mixed deciduous forest (Henderson & Harris, 1975).
[c] Data based on Ca budget from a *Liriodendron tulipifera* forest (Shugart et al., in press).
[d] Considers aboveground biomass pool. Cyclic renewal of structural roots (Kolesnikov, 1968) would lower turnover time. Tree mortality estimated from permanent plot resurvey (3 year interval) and likely underestimates the mortality rate over the duration of a forest generation.
[e] Turnover time based on available Ca and assumes all losses of Ca from soil are from the pool of available Ca.
[f] Total calculated as *sum* of elements in living and dead components of the ecosystem; element loss based on *sum* of all losses from ecosystem.

be bound in organic form to be utilized by the system. Since there are no large reservoirs of nitrogen in soil minerals, it might be anticipated that conservation mechanisms would be well-developed in terrestrial ecosystems.

In forest ecosystems carbon residence times of woody structure and soil organic matter range from 50 to 150 years (Table 1). The residence times of carbon and nitrogen, however, vary by 5 orders of magnitude amongcomponents of the system. These patterns are consistent with the potential role of accumulated organic matter as an alternate energy base. The small, but rapid fluxes of nitrogen through the decomposers are sufficient to maintain autotroph processes. The rapid activity of decomposers, resulting in a small pool of soluble nitrogen at any instant, serves to minimize leaching losses from the system. Because of organic matter accumulation in the ecosystem, the 1800 year mean residence time of nitrogen in the system is an order of magnitude greater than the residence time of nitrogen within any component of the ecosystem (Table 2). No single component or population within the system has a residence time sufficient to provide for nitrogen conservation. The long residence time of nitrogen is thus a unique ecosystem property.

Calcium cycling in a deciduous forest (Shugart et al., in press) illustrates the conservation of an element for which an abundant pedogenic source exists (Table 2). Calcium turnover in ecosystem components (available soil Ca, biomass, litter) is an order of magnitude more rapid than turnover of Ca in the total ecosystem. The interactions of autotrophs, big-slow components of the energy reservoir, and decomposers appear equally efficient at conserving minerals for which a significant source exists in primary minerals.

Rate regulation

Normal ecosystem function also includes the means of internal system regulation. Without rate regulation, the autotroph populations could exhaust available water and element resources resulting in collapse of the energy base. Similarly, utilization of the energy reservoir and remobilization of its element resources may require regulation of microbial processes. In an earlier paper (O'Neill et al., in press), we hypothesized that the complex food webs of ecosystems constitute a means of regulating ecosystem function. Wiegert & Owen (1971) and May (1973), among others, have discussed the stabilizing influence of heterotrophs on ecosystem processes. We consider the role of heterotroph regulation of ecosystem processes as essential to the persistence of ecosystems.

The magnitude (numbers times individual mass) of the heterotroph regulators in an ecosystem bears a relationship to the predominant type of energy base established in that system. For example, planktonic systems maintain an order of magnitude greater heterotroph biomass, e.g. 21 to 32 g dry matter m^{-2} (Riley, 1956; Harvey, 1950) than do forests e.g. 0.31 to 0.25 g m^{-2} (Satchell, 1971; Reichle et al., 1973a). Herbivore populations of small consumers in forests may exert more control than would be indicated by the small amount of organic matter actually consumed. While the heterotrophic consumption in forest canopies (in the absence of insect epidemics) may be only a few per cent of net primary production, the impact on photosynthetic potential can be more substantial (Reichle et al., 1973b). The significance of the role that consumers exert directly on the energy base may be small in comparison to the role that might be postulated for episodic outbreaks of populations. Surges in consumer populations maintain ecosystem heterogeneity and thus diversity of habitat and preservation of biotic components through elimination of locally dominant vegetation.

A characteristic of the energy reservoir of an ecosystem is its relative unavailability, with the utilization of this energy base and remobilization of element resources dependent upon the catabolic activities of microbes. High levels of microbial activity on detritus depend on processing of organic matter by fauna to expose greater surface area (van der Drift & Witkamp, 1960; Nef, 1957; McBrayer, 1973). Without comminution of organic detritus, microbial populations quickly exhaust the readily available substrate, remineralized elements are immobilized in senescing microbial colonies, and amounts of essential elements available for maintenance of the autotrophs become insufficient. The magnitude of the role of heterotrophs as regulators of energy and element flows is apparent from proportion of total ecosystem respiration relegated to heterotrophic activities. Heterotrophic respiration may account for between 35 and 55 per cent of total ecosystem respiration (autotrophic plus heterotrophic respiration). In a forest ecosystem, 95 per cent of total heterotrophic respiration is contributed by decomposers operating off the detritus energy base (Reichle et al., in press). Thus, it can be demonstrated that ecosystems allocate a substantial portion of their maintenance respiration energy to nutrient recycling. Consistent with the properties of a control system, however, regulation requires a smaller amount of energy than does operation of the system itself.

Ecosystem state variables and the biological environment

Four attributes of ecological systems, (i) establishment of an energy base, (ii) development of an energy reservoir, (iii) recycling of elements and, (iv) rate regulation, implicitly define the essential components or state variables essential to the function and persistence of the ecosystem. This set of state variables consists of a solar energy base, regulator organisms and an organic energy pool. At this level of organization the flow of energy and cycling of elements are inextricably interwoven; distinctions among trophic levels and species components are difficult to separate, since ecosystem function relies on mechanisms which are dependent on highly complex interactions among its trophic level components.

Ecosystem Metabolism

In the functional dynamics of ecosystems, energy is expended to facilitate material exchange. When energetic properties of ecosystems are categorized

Table 3. Comparative metabolic parameters of four contrasting ecosystems. All values in grams of carbon per square meter per year (from Reichle, 1975).

Property	Mesic Forest*	Xeric Forest†	Prairie‡	Tundra §
Gross Primary Production (GPP)	1620	1320	635	240
Autotrophic Respiration (R_A)	940	680	215	120
Net Primary Production (NPP)	680	600	420	120
Heterotrophic Respiration (R_H)	520	370	271	108
Net Ecosystem Production (NEP)	160	280	149	12
Ecosystem Respiration (R_E)	1470	1050	486	228
Production Efficiency [R_A/GPP]	0.58	0.52	0.34	0.50
Effective Production [NPP/GPP]	0.42	0.45	0.66	0.50
Maintenance Efficiency [R_A/NPP]	1.38	1.13	0.51	1.00
Respiration Allocation [R_H/R_A]	0.55	0.54	1.26	0.90
Ecosystem Productivity [NEP/GPP]	0.10	0.20	0.23	0.05

* Early successional deciduous forest (*Liriodendron tulipifera*) forest on alluvial soil (after Reichle et al., 1973a).

† *Quercus* and *Pinus* forest on sandy soil (after Woodwell & Botkin, 1970).

‡ US/IBP Grasslands Biome program, personal communication of raw data—calculations and interpretation by the author. Refinements after Andrews et al., (1974).

§ Adapted from the data of P. C. Miller, L. L. Tieszen, P. I. Coyne & J. J. Kelly in Bowen (ed.) 1972 Tundra Biome Symposium. Revisions of preliminary data after Tieszen and Coyne (personal communication).

according to nutritional processes of ecosystems instead of traditional trophic delineations, it is possible to compare the metabolic patterns of different ecosystems (Table 3). To illustrate the general patterns of energy flow in various ecosystems, metabolic parameters (ecosystem production equations) are summarized for four different ecosystems: a deciduous forest, a coniferous forest, a prairie and a tundra (Reichle, 1975). Total energy fixation (gross primary production) is correlated with both the standing crops and the length of the growing seasons Gross primary production is greatest in the temperate forest, followed by the prairie and finally in the tundra. Net primary production, which is gross primary production less autotrophic respiration, represents the yield of the energy-processing system. The costs of production, represented by autotrophic respiration, are dependent upon the mass of support tissues, but are also influenced by the mean annual temperature of the growing season. When expressed in terms of various ecosystem production efficiencies, it can be seen that the various ecosystems have adapted with comparable energy 'costs' for primary production.

When the production efficiency (R_A/GPP) of the ecosystems is compared, a striking similarity among all the ecosystems appears. Effective production (NPP/GPP) ratios suggest that those systems dominated by annual autotroph communities (prairie and tundra) without substantial energy reservoirs in perennial woody biomass have the highest yields relative to total carbon fixed. Ecosystem maintenance efficiency (R_A/NPP) illustrates how these systems allocate their carbon reserves between maintenance respiration and yield. Intuitively this allocation should be related to the quantity and persistence of non-photosynthetic biomass characteristic of each ecosystem. Maintenance metabolism is lowest in the prairie but progressively increases with the tundra and xeric forest, reaching maximum values in the mesic deciduous forest. The metabolic expenditures for heterotrophic respiration reflect the amount of net annual production that is recycled to maintain the available nutrient pools for continued production. This respiration allocation (R_H/R_A) illustrates that heterotrophic respiration is proportionally highest in the prairie ecosystem, intermediate in the tundra ecosystem, and lowest in the forest ecosystems. Ecosystem productivity (NEP/GPP), defined as the net carbon gained by the ecosystem relative to the gross primary production, is the energy cost of establishing the energy reservoir and the long term maintenance of element resources. Since appreciable NEP carbon increments can be added to soil organic matter as well as to standing crop, the range of these values between 5 and 23 per cent reflects site conditions and age of the system as much as it does any innate ecosystem property.

Thus, by analyzing the ecosystem in terms of the energy costs of material exchange, it becomes possible to establish a common framework for comparative ecosystem metabolism.

Ecosystem succession

An ecosystem must also be considered to be embedded in a biological environment. At the ecosystem level of organization this environment is composed, in part, of other ecosystems, just as an organisms's environment is composed of other organisms (Odum, 1969; Holling, 1973). This mosaic of ecosystems influences the biological material available to the system (MacArthur & Wilson, 1967). While a great deal of ecosystem heterogeneity can be the result of chance catastrophe (e.g. fire, wind and flooding), heterotrophs may play an unappreciated role in maintaining heterogeneity. Such a role is not in conflict with the general function of process control within the ecosystem. The episodic outbreak of heterotrophic regulators controls the availability of essential biological resources capable of colonizing openings resulting from stochastic climatic events.

In our concept of ecosystem function, biological succession is not considered as a fundamental principle of ecosystem function. The course of succession is ecosystem development; the end result of ecological succession is a state of maximum persistent biomass. Odum (1969) defines ecological succession in terms of three parameters: (1) succession is an orderly process of community development that is reasonably directional and, therefore, predictable, (2) succession results from modification of the physical environment by the ecosystem, and (3) succession culminates in a stabilized ecosystem in which maximum biomass and symbiotic function between organisms are maintained per unit of available energy flow. In Odum's view, succession involves increased control of, or homeostasis with, the physical environment directed towards some level of 'maximum protection from environmental perturbations'. Ecological succession in a heterogeneous environment is the logical result of the interactions among functional ecosystem processes. Increased homeostatic control of ecosystems during development, while buffering the biological environment from environmental fluctuations, ensures persistence through highly regulated long-term residence of energy pools and essential element resources.

Summary

Ecosystems are energy-processing systems, wherein vital ecological processes such as nutrient cycling are energy-consuming operations. Analytical efforts aimed at deducing the functional optimization of ecosystems, therefore, must evaluate the energy 'costs' of system function in the context of the processes internal to the ecosystem. Such an approach to ecosystem analysis requires conceptual formulation of the system as a hierarchy of interacting subsystems. Unique system properties arise from the structural organization and interaction of these subsystem components.

Ecosystems, regardless of type or complexity, all appear to exhibit common properties of persistence and growth. Ecosystems tend to display consistent patterns of optimization towards maximum persistent biomass. The specific nature of ecological mechanisms underlying this phenomenon are complex, but all ecosystems have evolved regulative mechanisms whereby a secure energy base is established in the presence of a fluctuating environment. Autotrophic components of ecosystems display one of two basic properties (or combinations thereof): (i) small individuals with rapid turnover and (ii) large individuals with slow turnover. Combinations of these two properties serve to provide homeostatic mechanisms with attributes of both rapid response and long-term stability in the photosynthetic base for energy conversion.

Concomitantly, all ecosystems have also developed mechanisms for energy storage as an operational basis for maintaining homeostasis. This reservoir energy base is characteristically large, with slow response time, but is critical for nutrient storage and remobilization. Many ecosystems representative of a wide range of structural and climatic conditions possess such an energy reservoir in organic detritus. Patterns in energy metabolism and carbon allocation for a variety of ecosystem types have been reviewed which show similar characteristics of net ecosystem production, i.e. energy storage in detritus. This storage represents the energetic costs of maintaining essential elements in a form which can be readily utilized by the ecosystem.

Thus, any concepts of ecosystem function must recognize that ecosystems are energy-processing units regulated by, and inextricably connected to, the cycles of nutrient elements in the system. Through internal system properties achieved by hierarchical organization of trophic components and populations, ecosystems achieve stability and persistence by establishing secure bases of both energy fixation and energy storage.

References

Andrews, R., D. C. Coleman, J. E. Ellis and J. S. Singh. 1974. Energy flow relationships in a shortgrass prairie ecosystem. Proc. 1st Int. Congr. Ecol. Pudoc, Wageningen.

Bowen, S. (ed.). 1972. Proceedings 1972 Tundra Biome Symposium. U.S. IBP Tundra Biome Report. CRREL U.S.A.

Drift, J. van der and M. Witkamp. 1960. The significance of the breakdown of oak litter by *Enoicyla pusilla* Burm. Arch. Neerl. de Zoöl. 13: 486–492.

Duvigneaud, P. and S. Denaeyer-DeSmet. 1970. Biological cycling of minerals in temperate deciduous forests. In: Analysis of Temperate Forest Ecosystems. D. E. Reichle, ed. 199–225.

Gore, A. J. P., and J. S. Olson. 1967. Preliminary models for accumulation of organic matter in an *Eriophorum/Calluna* ecosystem. Aquilo, Ser. Botanica, 6: 297–313.

Harley, J. L. 1959. Biology of Mycorrhiza. Leonard Hill, London.

Harris, W. F., P. Sollins, N. T. Edwards, B. E. Dinger and H. H. Shugart. 1972. Analysis of carbon flow and productivity in a temperate deciduous forest ecosystem. In: IBP V General Assembly Symposium, 'Productivity of World Ecosystems.' D. E. Reichle, D. Goodall & P. Baker, eds. (in press).

Harris, W. F., P. Sollins, N. T. Edwards, B. E. Dinger, and H. H. Shugart. 1975. Analysis of carbon flow and productivity in a temperate deciduous forest ecosystem. In: 'Productivity of World Ecosystems,'IBP V General Assembly Symposium, Seattle, Sept. 1972. D. E. Reichle, J. Franklin, D. Goodall & P. Baker (eds.). National Academy of Sciences.

Harvey, H. W. 1950. On the production of living matter in the sea off Plymouth. J. Mar. Biol. Assoc. U.K. 29: 97–137.

Heatwole, H. and R. Levins. 1972. Trophic structure stability and faunal change during recolonization. Ecology 53: 531–534.

Henderson, G. S. and W. F. Harris. 1975. An ecosystem approach to characterization to the nitrogen cycle in a deciduous forest watershed. In: Forest Soils and Forest Land Management. B. Bernier & C. H. Wiarget, eds.

Holling, C. S. 1973. Resilience and stability of ecological systems. Ann. Rev. Ecol. Syst. 4: 1–23.

Kolesnikov, V. A. 1968. Cyclic renewal of root in fruit plants. In: International Symposium, USSR. Methods of productivity in root systems and rhizosphere organisms: Reprinted for the IBP by Biddles. Ltd., Guildford, U.K. 102–106.

MacArthur, R. A. and E. O. Wilson. 1967. The theory of island biogeography. Princeton University Press, Princeton, N.J.

May, R. M. 1973. Stability and complexity in model ecosystems. Princeton University Press, Princeton, N.J.

McBrayer, J. F. 1973. Energy flow and nutrient cycling in a cryptozoan food web. EDFB-IBP 73-8, Oak Ridge National Laboratory, Oak Ridge, Tennessee.

McGinnis, J. T., F. B. Golley, R. G. Clements, G. I. Child, and M. J. Duever. 1969. Elemental and hydrologic budgets of the Panamanian tropical moist forest. Bio-Science 19: 697–700.

Milsum, J. H. 1966. Biological Control Systems Analysis. McGraw-Hill, New York.

Monk, C. L. 1966. An ecological significance of evergreenness. Ecology 47: 504–505.
Nef, L. 1957. État actuel des connaissances sur le rôle des animaux dans la decomposition des litiers de forést. Agricultura 5: 245–316.
Odum, E. P. 1969. The strategy of ecosystem development. Science 164: 262–270.
Odum, E. P. 1971. Fundamentals of Ecology (3rd ed.). W. B. Saunders & Co., Philadelphia.
Odum, H. T., & R. F. Pigeon (eds.). 1970. A tropical rain forest. Division of Technical Information, U.S. Atomic Energy Commission, Oak Ridge, Tenn. 1650 pp.
Olson, J. S. 1965. Equations for cesium transfer in a *Liriodendron* forest. Health Physics 11: 1385–1392.
O'Neill, R. V., W. F. Harris, B. S. Ausmus and D. E. Reichle. 1974. A theoretical basis for ecosystem analysis with particular reference to element cycling. In: Symposium Proceedings, Mineral Cycling in the Southeastern United States AEC-CONF. (in press).
Patten, B. C. 1971. A primer for ecological modeling and simulation with analog and digital computers. In: Systems Analysis and Simulation in Ecology, Volume I. Academic Press, New York.
Reichle, D. E. 1975. Advances in Ecosystem Analysis, BioScience 25: 257–264.
Reichle, D. E., B. E. Dinger, N. T. Edwards, W. F. Harris and P. Sollins. 1973. Carbon flow and storage in a woodland ecosystem. In: Carbon and the Biosphere. G. M. Woodwell & E. Pecan, eds. USAEC-CONF-720510, Washington, D. C. 345–365.
Reichle, D. E., R. A. Goldstein, R. I. van Hook and G. J. Dodson 1973b. Analysis of insect consumption in a forest canopy. Ecology 54: 1076–1084.
Reichle, D. E., J. F. McBrayer and B.S. Ausmus. 1975. Ecological energetics of decomposers in a deciduous forest. In: Proc. Vth Internat. Colloquia of Soil Zoology. Prague, Czechoslovakia, October 1973.
Riley, G. A. 1956. Oceanography of Long Island Sound, 1952–54. IX. Production and utilization of organic matter. Bull. Bingham Oceanogr. Coll. 15: 324–344.
Satchell, J. E. 1971. Feasibility study of an energy budget for Meathop Wood. In: Productivity of Forest Ecosystems. P. Duvigneaud, ed. UNESCO, Paris. 619–630.
Shugart, H. H., D. E. Reichle, N. T. Edwards and J. R. Kercher. A model of calcium-cycling in an east Tennessee *Liriodendron* forest: Model structure, parameters and analysis in the frequency domain. (Unpublished maunscript, submitted to Ecology).
Smith, F. E. 1970. Analysis of ecosystems. In: Analysis of Temperate Forest Ecosystems. D. E. Reichle, ed. Springer Verlag, New York and Heidelberg. 7–18.
Taylor, F. G., Jr. 1974. Phenodynamics of production in a mesic deciduous forest. In: Phenological and Seasonality Modeling. H. Lieth & F. Stearns, eds. Springer Verlag, Heidelberg, Berlin, New York. 237–254.
Watt, K. E. F. 1966. The nature of systems analysis. pp. 1–14 In: Systems Analysis in Ecology, N. E. F. Watt, (ed.). Academic Press, N.Y. 276 pp.
Weiss, P. A. 1971. The basic concept of hierarchic systems. pp. 1–44. In: Hierarchically Organized Systems in Theory and Practice, P. A. Weiss, (ed.). Hafner, New York. 263 pp.
Wells, C. G., and J. R. Jorgensen. 1975. Nutrient cycling in loblolly pine plantations. In: Proc. 4th North American Forest Soils Conf., Quebec, Aug. 20–24, 1973.
White, H. J., and S. Tauber. 1969. Systems analysis. Saunders, Philadelphia. 499 pp.
Whitfield, D. W. A. 1972. Systems analysis. pp. 392–409. In: L. C. Bliss (ed.), Devon Island IBP Project, High arctic ecosystem. Department of Botany, University of Alberta. 413 pp.
Wiegert, R. G., and D. F. Owen. 1971. Trophic structure, available resources and population density in terrestrial versus aquatic ecosystems. J. Theor. Biol. 30: 69–81.

Woodwell, G. M., and D. B. Botkin. 1970. Metabolism of terrestrial ecosystems by gas exchange techniques, pp. 73–85. In: Analysis of Temperate Forest Ecosystems D. E. Reichle, (ed.). Springer-Verlag, Berlin-Heidelberg-New York. 304 pp.

Author's addresses:
D. E. Reichle
Environmental Sciences Division
Oak Ridge National Laboratory
Oak Ridge, Tennessee
USA
R. V. O'Neill
Environmental Sciences Division
Oak Ridge National Laboratory
Oak Ridge, Tennessee
USA
W. F. Harris
Environmental Sciences Division
Oak Ridge National Laboratory
Oak Ridge, Tennessee
USA

The role of bacteria in energy flow and nutrient cycling

H. Veldkamp

Introduction

Both in terrestrial and aquatic environments a considerable part of organic matter originating in primary production ends up in one form or another in what is represented in flow diagrams as a box with inscription 'decomposers'. For instance, an analysis of the Serengeti Plains (Phillipson, 1973) shows that of a net primary production of 300 g/m^2/year, no less than 68 % is broken down by decomposers. The fate of organic substrates broken down has been most extensively studied in bacteria. And as this group of organisms is also by far the most versatile metabolically, the following discussion will be restricted to bacteria.

Types of energy metabolism in bacteria

In bacteria three main types of energy metabolism occur. Firstly, photosynthetic bacteria are able to convert light energy into chemically bound energy. As bacterial photosynthesis is restricted to anaerobic environments, its contribution to primary production and nutrient cycling is only of local importance (Brock, 1974).

A second group, the chemolithotrophic bacteria, derives energy from the aerobic oxidation of inorganic substrates, such as Fe^{2+}, S^{2-}, S, NH_4^+ and NO_2^-.

Many of these processes are marginal from an energetic point of view (Zajic, 1969; Brock, 1974). This means that very large numbers of inorganic ions have to be oxidized to produce enough ATP for the synthesis of a new bacterial cell from simple ingredients (CO_2 and some minerals). These organisms therefore play an important role in nutrient cycling in all aerobic habitats, but their 'standing crop' is low.

A third type of energy metabolism is present in chemoorganotrophic bacteria, which derive energy from the oxidation of organic compounds. They can perform one (some more) of three energy generating processes: (1) *aerobic respiration* (O_2 accepts electrons from oxidized organic substrate), (2) *anaerobic respiration* (inorganic compounds other than O_2 accept the electrons under anaerobic conditions; e.g. NO_3^-, SO_4^{2-}, CO_2) and (3) *fermentation* (organic substrate is oxidized anaerobically; intermediary metabolism provides organic compounds to be reduced).

The amount of ATP produced in the oxidation of a molecule such as glucose decreases in the above order (cf. Brock, 1974). However, the amount of ATP needed to synthesize a new bacterial cell from glucose as the only C- and energy source, is approximately the same in the three groups of organisms. This thus means that the amount of biomass that can be formed per mole of glucose broken down depends in the first place on the kind of energy metabolism involved (cf. Stouthamer & Bettenhausen, 1973; Stouthamer, 1975). The three

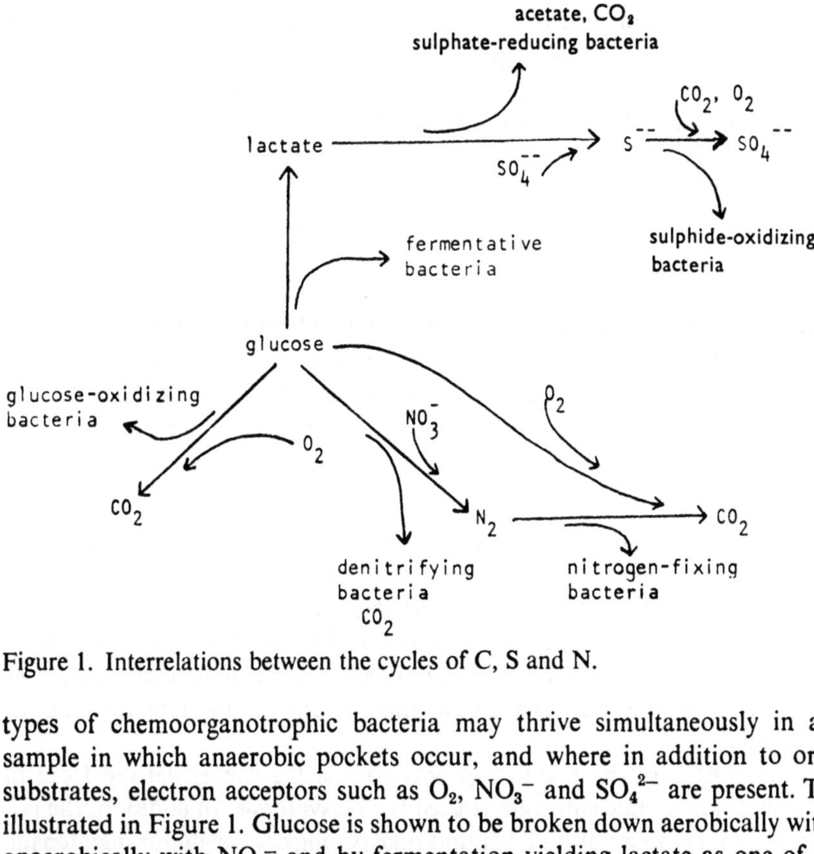

Figure 1. Interrelations between the cycles of C, S and N.

types of chemoorganotrophic bacteria may thrive simultaneously in a soil sample in which anaerobic pockets occur, and where in addition to organic substrates, electron acceptors such as O_2, NO_3^- and SO_4^{2-} are present. This is illustrated in Figure 1. Glucose is shown to be broken down aerobically with O_2, anaerobically with NO_3^- and by fermentation yielding lactate as one of many possible fermentation products. Figure 1 also shows that cycles of the elements cannot be considered as separate entities. Lactate, derived from glucose fermentation, can be used in anaerobic respiration of sulphate reducing bacteria, forming sulphide. This in turn serves as a substrate for chemolithotrophic bacteria when oxygen is present. Therefore, part of the chemical energy present in glucose may end up in sulphide, which can consequently be oxidized to sulphate. Similar considerations hold for interrelations between the cycles of C and N. Therefore, ideally, the cycling of the elements should be studied simultaneously.

Nutrient limitation as a factor in biomass formation in chemoorganotrophic bacteria

It is often thought that in the breakdown of organic substrates there is a certain fixed ratio between substrate-C ending up in cell material and in metabolic end products, such as CO_2. This concept arose from laboratory experiments in which bacterial populations were growing at maximal rate in batch culture (closed culture system) in the presence of excess of all nutrients needed. However, in natural environments bacteria generally grow at submaximal rate, determined by the rate at which an essential nutrient becomes available. Growth-rate limitation can be mimicked in the laboratory by making use of an open culture system, the chemostat (Veldkamp & Kuenen, 1973; Jannasch & Mateles, 1974; Veldkamp, 1975). And such studies have shown that growth yield is highly dependent on growth rate (Stouthamer & Bettenhausen, 1973; Stouthamer, 1975; Veldkamp, 1975). Without any doubt, this will also be the case in bacteria growing in natural environments.

The specific growth rate (μ) of a bacterial population is defined as

$$\mu = 1/x \cdot dx/dt = d \ln x/dt = \ln 2x/\ln x \cdot 1/t_d = \ln 2/t_d \qquad (1)$$

where x is cell concentration (mg dry weight/ml) at time t, and t_d is time required for the concentration of organisms to double. The relation between the specific growth rate of a bacterial population and the concentration of a growth rate-limiting substrate is described empirically by the equation (Monod, 1942; 1950):

$$\mu = \mu_{max} (s/K_s + s) \qquad (2)$$

where μ = specific growth rate, μ_{max} is maximum specific growth rate, s is substrate concentration and K_s is a saturation constant, numerically equal to the substrate concentration at 0.5 μ_{max}. A graphical presentation of this relation is shown in Figure 2. Note that small changes in the concentration of the growth rate-limiting substrate, when this is low, have a considerable effect on growth rate. And, as will be described below, this in turn causes marked changes in growth yield.

In a growing bacterial population only part of the energy derived from the oxidation of an organic substrate is used for biosynthesis. In addition, the cells need a constant amount of ATP per unit time per mg dry weight for maintenance purposes (Pirt, 1965; Stouthamer & Bettenhausen, 1973; Stouthamer, 1975). The relation between specific growth rate and growth yield is given by the following equation (Schulze & Lipe, 1964; cf. van Uden, 1969):

$$q = \mu/Y = \mu/Y_G + m_s \qquad (3)$$

where q is specific substrate uptake rate (g of substrate used/g dry weight cells/h); the observed growth yield $Y = g$ dry weight/g of substrate used; Y_G = the growth yield that would have been obtained if there were no maintenance requirement; the maintenance coefficient $m_s = g$ substrate/g dry weight/h; μ is

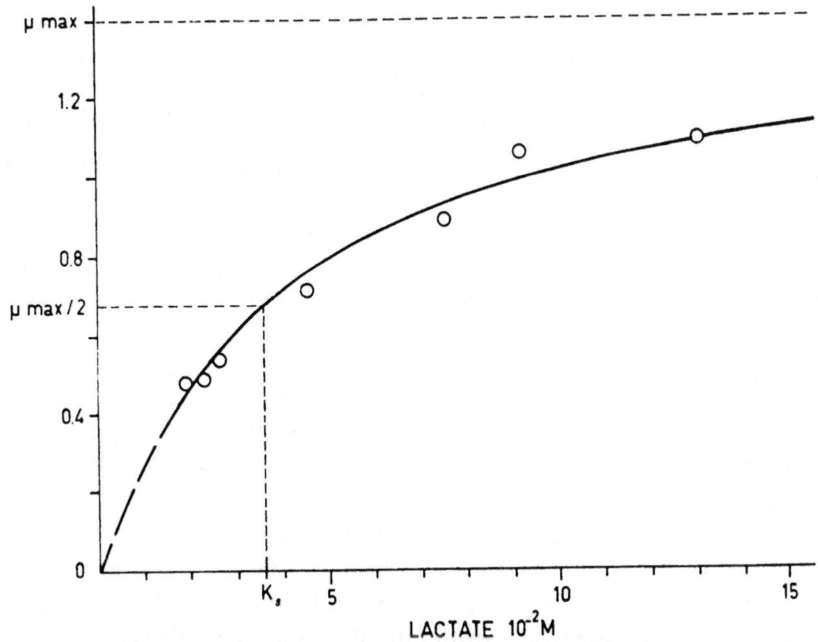

Figure 2. Relation between specific growth rate (μ) and concentration of growth rate-limiting substrate (lactate) in *Propionibacterium shermanii* (Jerusalimsky, 1967).

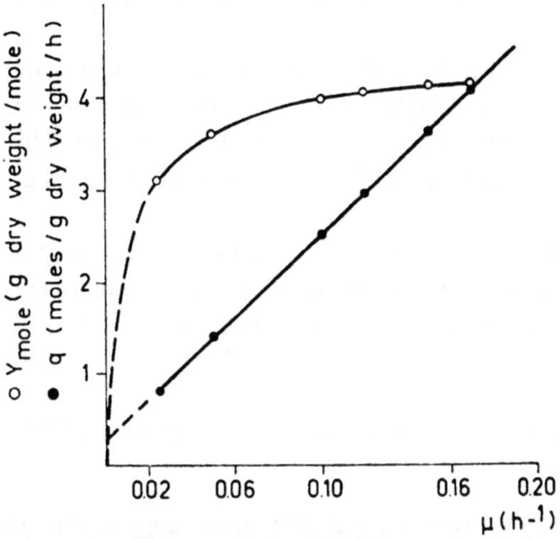

Figure 3. Molar growth yield (Y_{mole}: *g* cell dry weight formed per mole of oxalate consumed) and specific oxalate uptake rate (q) as a function of specific growth rate (μ). Chemostat data obtained with *Pseudomonas oxalaticus* grown in a culture medium with excess minerals and oxalate, the only C- and energy source, as growth rate-limiting substrate (Harder, 1974).

specific growth rate in h^{-1}. In this equation μ/Y_G represents substrate consumption for growth, and m_s the consumption for purposes other than growth, collectively called maintenance purposes.

Figure 3 shows a typical example of q and Y as a function of μ. Extrapolation of q to zero growth rate gives an intercept with the ordinate, which represents the growth rate-independent maintenance requirement (m_s). The slope of the line is $1/Y_G$. Y can be calculated from equation 3, once m_s and Y_G have been established. When Y is measured directly, it appears that experimental and predicted values are in good agreement.

When a bacterial population is grown in a glucose-limited chemostat (aerobically; mineral medium with glucose as only C- and energy source), realistic values for Y_G and m_s are 0.5 and 0.05, respectively (Pirt, 1965). Equation 3 then shows that at specific growth rates of 1 and 0.01 h^{-1}, the corresponding cell yields (Y) are 0.48 and 0.14.

In terms of biomass formation, the most efficient use is made of the C- and energy source when this limits growth rate. The growth yield (Y) with respect to C- and energy source is lower in populations limited by such nutrients as HPO_4^- or NH_4^+. The lower the growth rate, the larger the difference between yields of C- limited populations and populations limited by another nutrient.

Thus, whatever the growth rate-limiting substrate, activities of chemo-organotrophic bacteria shift from biomass formation to nutrient cycling with decreasing growth rates. And the effect is less pronounced in C- limited populations.

Predation should favour biomass formation. Provided the rate at which the growth rate-limiting substrate becomes available remains constant, introduction of a predator results in an increase in the concentration of this substrate. This results in an increase in specific growth rate (μ in Fig. 2), which in turn gives an increase in growth yield (Y in Fig. 3).

In summary, the amount of bacterial biomass that can be formed per mole of organic substrate consumed depends on: the type of energy metabolism involved, growth rate, kind of substrate limitation, and degree of predation.

References

Brock, T. D. 1974. Biology of microorganisms. Prentice-Hall Inc., Englewood Cliffs, New Jersey.

Harder, W. 1974. Unpublished.

Jannasch, H. W. and R. I. Mateles. 1974. Experimental bacterial ecology studied in continuous culture. Adv. Microbial Physiol. 11: 165–208.

Jerusalimsky, N. D. 1967. Bottle-necks in metabolism as growth rate controlling factors. In: Microbial physiology and continuous culture, E. O. Powell, C. G. T. Evans, R. E. Strange & D. W. Tempest. eds. pp. 23–33. H.M.S.O., London.

Monod, J. 1942. Recherches sur la croissance des cultures bactériennes. Hermann et Cie, Paris.

Monod, J. 1950. La technique de culture continue: théorie et applications. Annls Inst Pasteur. 79: 390.

Phillipson, J. 1973. The biological efficiency of protein production by grazing and other land-based systems. In: The biological efficiency of protein production, J. G. W. Jones, ed. pp. 217–237. Cambridge University Press, London.
Pirt, S. J. 1965. The maintenance energy of bacteria in growing cultures. Proc. Royal Soc. B. 163: 224–231.
Schulze, K. L. and R. S. Lipe. 1964. Relationship between substrate concentration, growth rate, and respiration rate of *Escherichia coli* in continuous culture. Arch. Mikrobiol. 48: 1–20.
Stouthamer, A. H. 1975. Yield studies in microorganisms. Merrow Publ. Co. Ltd, Watford, Herts. (In Press).
Stouthamer, A. H. and C. Bettenhausen. 1973. Utilization of energy for growth and maintenance in continuous and batch cultures of microorganisms. Biochim. Biophys. Acta 301: 53–70.
Uden, N. van 1969. Kinetics of nutrient-limited growth. Ann. Rev. Microbiol. 23: 473–487.
Veldkamp, H. 1975. Continuous culture in microbial physiology and ecology. Merrow Publ. Co. Ltd, Watford, Herts. (In press).
Veldkamp, H. and J. G. Kuenen. 1973. The chemostat as a model system for ecological studies. In: Modern methods in the study of microbial ecology, T. Rosswall, ed. Proceedings Symposium. Uppsala 1972. pp. 347–357. Swedish National Research Council, Stockholm.
Zajic, J. E. 1969. Microbial biogeochemistry. Academic Press, New York.

Author's address:

H. Veldkamp
Laboratorium voor Microbiologie
Biologisch Centrum
Haren (Gr.)
The Netherlands

The flow of energy and matter between trophic levels (with special reference to the higher levels)

S. S. Schwarz

Introduction

The interrelationships between producers and consumers are determined by their numbers, by the effectiveness with which energy is harnessed by the lower trophic levels, by the speed of renewal of dominant populations, by the ability of producers to renew consumed production, and by the relationship between energy required for maintainance and that available for production in the dominant species of different food chains. Such interrelationships determine the productivity and stability of the biogeocoenoses (BGC) and the peculiarities of energy and matter transfer. International Biological Programme investigations published recently have provided many data on the structure of BGC. Special attention is therefore paid in this paper to the functional characteristics of ecosystems.

Differences between biomes

Even rough calculations show that there are fundamental differences in biogeocoenotic indices between various biomes. In tundra the standing crop of the plants exceeds that of the zoomass by 15 times, in meadow steppes by 40 times, in cereal steppes by 50 times, in deciduous forests by 300 times, and in coniferous forests by 1200 times. The annual production of the phytomass compared with that of the zoomass also shows great differences: this index is 3 for deciduous forest, 5 for cereal steppes, 15 for meadow steppes, and 20 for coniferous forests (Isakov and Panfilov, 1969).

The consumer structure is also different. The proportions of saprophages: phytophages: predators is 20:2:1 in coniferous forests, 250:30:1 in oak woodlands, 20:5:1 in dry steppes. These rough calculations, based on data derived from various sources, demonstrate clearly the distinctions in character and the influence of biogeocoenotic processes on different trophic levels and in different biomes. The structure and function of the BGC inevitably changes sharply with alteration of the dominant species. For instance, in tundra the total biomass of phytophages exceeds that of the predators by about 10 times during quiescent years and more than 1000 times during lemming population peaks. Everyone is familiar with local changes in biogeocoenotic structure. Despite the relatively low productive processes in tundra, there are regions of *Calamagrostis* herb associations that produce a crop of up to 300–800 kg/ha of unconsumed green material during a two month growing season.

Great differences are also revealed in the structure of geographically and functionally related BGC. This can be illustrated by the biomass proportions of invertebrates in tundra with different types of ground (Fig. 1). The harmonious development of the interacting BGC of different structures is maintained by special mechanisms of coadaptation between the production processes in different ecosystems within the same BGC.

In this connection let us analyse the development of an amphibian. The biomass of eggs laid by a population of *Rana arvalis* in a given lake coincides with the biomass of young frogs leaving the water (Fig. 2). Of course, some divergences from this general rule may be observed, but there can be no doubt that this has an important role in the coordination of terrestial and aquatic subsystems of one BGC. For instance, the zoomass of arthropods inhabiting shallow basins in tundra (20% of the investigated territory) approaches 70 kg/ha (N. Olshwang, *in litt.*); up to 1.5 kg/ha of the zoomass is withdrawn from ponds by insects completing their metamorphosis, and this represents 20% of the zoomass of the terrestrial tundra BGC. Even when terrestrial ecosystems predominate, their biogeocenotic structure is determined by conditions in aquatic ecosystems where the larvae of the dominant species develop. For example, the biomass of larval *Aedes communis* mosquitoes in forest ponds was calculated to be 17 kg/ha; the imagines withdraw some 16 kg/ha from the ponds (which, if distributed randomly, would add approximately 1 kg/ha zoomass to the terrestial ecosystem, N. Nikolaeva, *in litt.*).

The development of species in adjacent ecosystems must naturally be well coordinated. Otherwise a sharp increase of a certain phase developed at the expense of one system might cause disturbances in another one characterized by a tighter energy budget. Mechanisms regulating this process revealed in *Rana arvalis* merit careful study.

The influence of population fluctuations

Population mechanisms regulating biogeocenotic processes play a still greater role when the numbers of one of the dominant species increase sharply. Let us analyse lemming peaks. In the autumn-winter 1972/73 the population density of *Dicrostonyx torquatus* in the moss-grass tundra reached 1000 holes per hectare. Towards summer 1973 some 70% of the dominant plant species (*Carex globularis*, *Eriophorum polystachyon*, *E. russedum*) had been gnawed. By this time, however, the lemming density had fallen sharply as a result of higher mortality and a lower breeding rate, and the biomass of cotton grass regenerated to 90% of the potential biomass by July, while that of the sedges was up to 40%, the number of shoots being almost completely renewed on separate plants. The number of lemmings fell sharply though the stock of food was practically

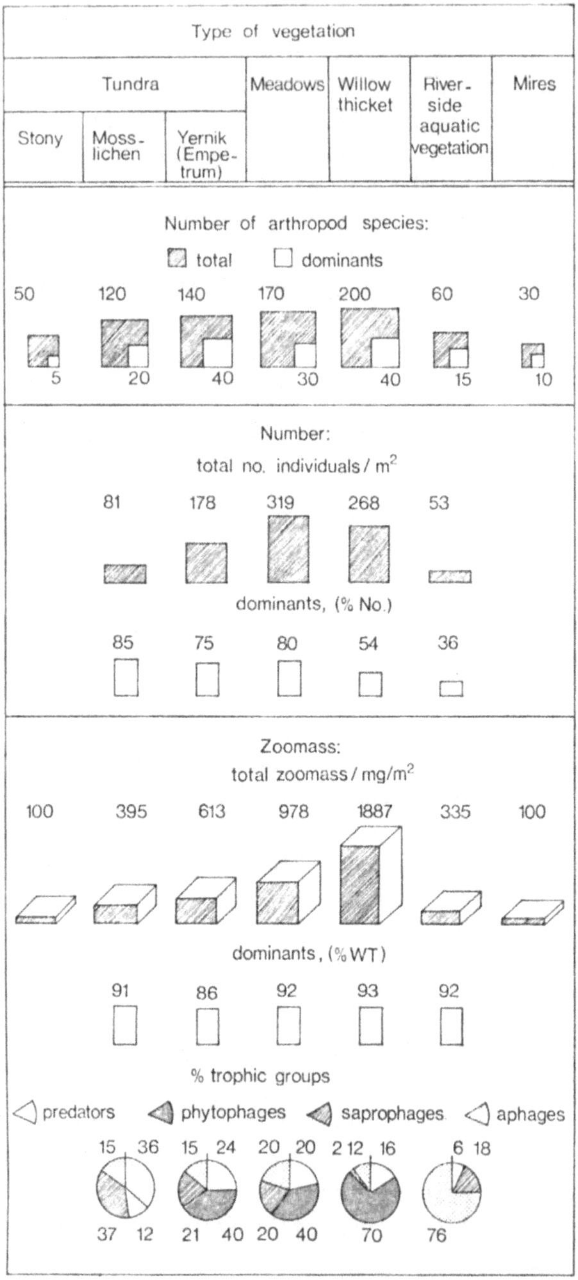

Figure 1. Biotopic spectrum of arthropods in the Polar Urals.

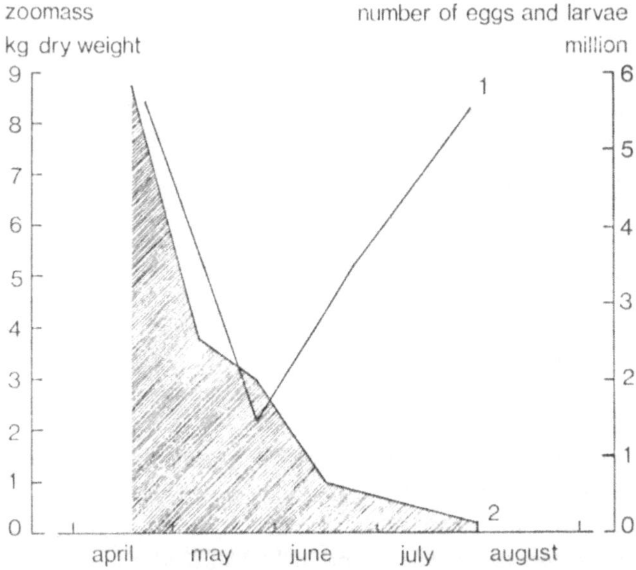

Figure 2. The relationship between zoomass (curve 1) and numbers (curve 2) of *Rana arvalis* during metamorphosis in a 1.5 ha pond in the South Urals, 1968 (after Shchupak, 1970). The overall loss of zoomass from the aquatic system was 1.2 %, compared with a mortality of 98.5 %.

inexhaustible. This example shows that in natural ecosystems (not disturbed by man) even mass breedings of dominant species do not disturb the BGC balance. Furthermore, there are grounds to assert that the mass reproduction of separate species, for a long time considered to upset the balance of nature, is in reality a necessary factor in preserving the BGC productivity. In this example, the peak and crash of the lemming population caused an enrichment of the soil with organic matter and minerals. Calculations showed that no less than 300 kg/ha of organic matter assimilated by the plants originated from carcasses during the mass breedings of lemmings in the tundra, and a considerable amount of minerals (circa 1000 to 1500 kg/ha) came from the excretions of the rodents, viz: Si 6.0; Al 5.0; Fe 0.5; P 4.0; Ca 6.0; Mg 3.0; Mn 0.2; K 15.0, Na 1.8; S 1.0 kg/ha. These figures are very approximate. They were estimated from observations on the number of lemmings and on their food preferences to elucidate the chemical composition of wastes (information from various unpublished works). Some mistakes in these calculations may have arisen from inevitable mistakes in estimating the energetic and mineral turnover of rodents in nature. Nevertheless, when comparing these figures with the chemical composition of grass litter in the tundra (Table 1) it becomes apparent that population peaks do not disturb the BGC; moreover, they become an important factor in the maintenance of the BGC balance.

Table 1. The chemical composition of a grass litter of sedges and cotton grass (% of dry weight)

Element	Sedges	Cotton grasses
Si	0.03	0.25
Al	0.05	0.18
Fe	<0.01	0.03
P	0.08	0.20
Ca	0.80	0.20
Mg	0.40	0.16
Mn	0.13	0.03
K	0.60	2.40
Na	0.01	0.07
S	0.33	0.15

Similar data have been obtained by observations on insects. In the fourth year of mass reproduction of gypsy moth (*Porthetria dispar*) the amount of organic matter introduced into the soil may be as much as 5.5 t/ha, exceeding the weight of grass litter (Rafes, 1962). The temporarily disturbed phytocenotic balance is quickly re-established.

It should be emphasized that the phytocoenotic response to the population peaks of consumers is not, as it is commonly assumed, so vastly different from the biogeocoenotic processes during 'calm years'. To keep their numbers at a constant, but not high, level, many dominant species undergo a geometric increase in reproduction accompanied by an equally high mortality. In the subarctic region a pair of tundra voles *Microtus oeconomus* produces no less than four generations of voles. Only animals from the last generation survive the autumn and form a winter population. But most of these animals die off during the long Arctic winter. The offspring of only one reproducing pair consume no less than 400 kg/ha/yr of phytomass. It becomes apparent that to keep its population density at a constant, and not high, level (c. 5 pairs per ha) a population of voles consumes some 400 kg/ha of plant material, which corresponds to a high level of grass stand in northern regions. This results from the high mortality of the animals.

This conclusion was supported by field observations on *Microtus agrestis*; Figure 3 shows the total number of voles descended from a single pair, as determined by total marking. During these observations, which allowed the fates of individual animals to be traced (their birth, development and death) it was estimated that to keep the population stable one reproducing pair consumed 265 kg/ha phytomass of grass-bean-herb meadow, including no less than 1.5 kg of phosphate and 15 kg of potassium. The phytocoenosis retained its normal productivity and structure, in spite of the continuous consumption of plants produced, by the effective utilization of nutrients which are prepared for rapid consumption by the activity of the animals.

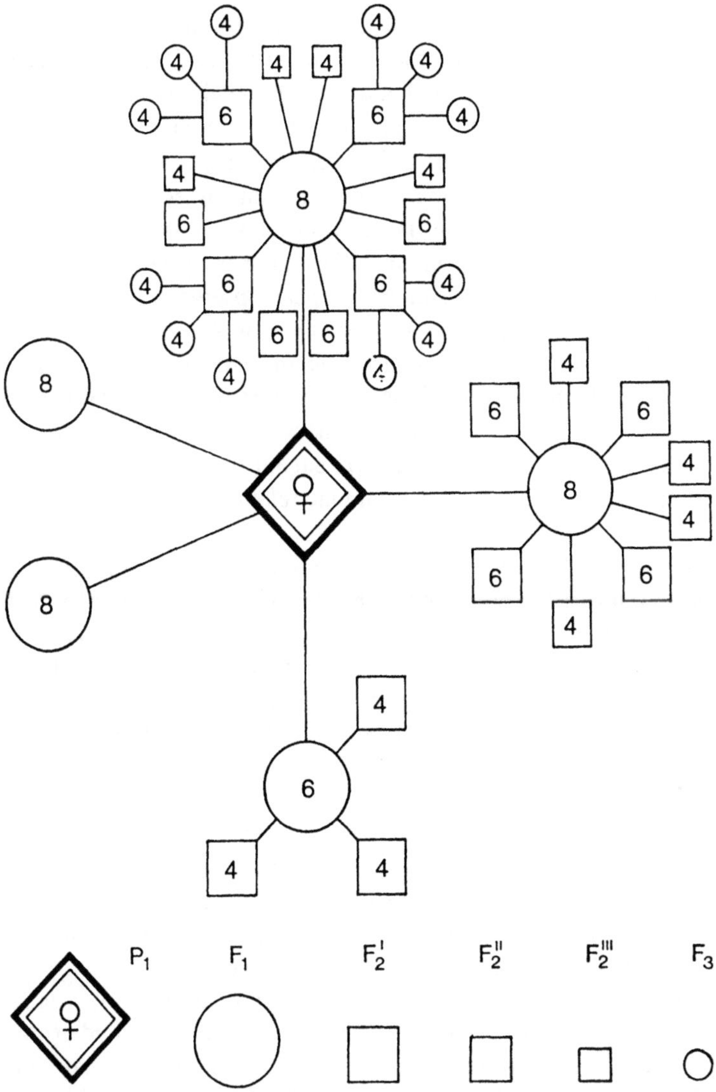

Figure 3. Reproduction in *Microtus agrestis*: P_1 overwintered female, F_1 its offspring; F_2^{i} F_2^{ii}, F_2^{iii}, the first, second and third litters of the F_2 generation; F_3 offspring of the first F_2 litter. Figures inside symbols indicate the number of animals in the litter. Mean litter size was 6.2. The total number of voles in the progeny was 196, with a sex ratio of 1:1.

The influence of the consumers on primary production

The essence of the population mechanisms regulating biogeocoenotic processes is based on the general fact that the population peak is generally followed by the population crash before the equilibrium of the BGC is disturbed on different trophic levels. Naturally these mechanisms play an important role in the life of the BGC only with regard to dominant species. Therefore it is necessary to show that a typical BGC structure is characterized by a small number of dominant species, which form the main basis of the trophic levels, and satellite species whose function is to ensure the normal activity of the dominants (Fig. 1). Irrespective of great differences in the total number and biomass of insects in various types of tundra BGC, and of differences in proportions of species belonging to different trophic levels, the role of dominants in creating biological production appears to be practically the same. Many similar examples could be given. It is important to stress, however, that the dominant producer in many BGC lies outside the control of the consumers. One can observe this on subarctic willows. In the Yamal peninsula these are damaged by 42 species of insect, especially by the sawflies *Amauronematus harpicola* and *Phytodecta pallidus* and by a leaf-eating beetle. The biomass of these species is up to 320 mg (80 mg dry weight) per m^2 (80–90% of the total insect biomass). The productivity of the willows ranges from 24 to 235 g/m^2; consumers eat no more than 7% of their potential food supply (Bogachova, *in litt.*). Similar relationships are typical for young BGC.

To evaluate such data the characteristic response of the plant to damage needs to be taken into account. It is known that even during mass breedings of leaf-eating insects the weight of leaves changes little for a long time because the hyperfunction of the chloroplasts compensates for this. In cases where losses exceed the ability of the compensating reactions, the decrease in leaf weight exceeding the weight of consumed food points to a pathological process. There is convincing evidence that reactions are specific for different species in the producer-consumer system. When the population of specialized consumers increases, the productivity of the foodplants decreases, but the productivity of other producers rises. Following a four-year increase in the number of the moth *Tortrix viridana* the accretion of oak declined by 21%, but that of the ash rose by 58%, and the total increment of the first trophic layer remained unchanged (Rafes, 1968). Similar phenomena have been noticed when sharp changes in population were caused by changes in climate. Numerous investigations made by N. N. Danilov in the southern tundra have shown that within related groups (*Anthus*, pipits, for example) northern species stay to nest if spring is late, but mostly southern ones nest here if spring comes early. On the whole the total number of animals playing similar roles in the ecosystem varies little. A change of dominant species within one trophic level is of great importance. The specific role of a particular species should not be underestimated, but neither should it be exaggerated when the question concerns general biological characteristics. Five amphibian species were studied in connection with this problem (*Rana arvalis*, *R. temporaria*, *R. macrocnemis*, *R. camerani*, *Pelobates syriacus*). It was found that the tadpoles of all these species utilized their food with almost equal efficiency (9.0 to 13.1% of the food consumed being converted into animal biomass), despite great differences in growth and development rates when these were regulated experimentally.

This type of information permits the evaluation of the intensity of energy balances on different trophic levels without concrete measurements.

Differences between the actions of homeotherm and poikilotherm consumers

Animals of entirely different metabolic types play different roles in the BGC. This may be proved by comparing a frog and a bird of about the same body weight and food regime. The activity of the homeotherm animal is impaired if the quantity of prey is unlimited but its population density is low, because this makes its search for prey energetically disadvantageous. A poikilotherm frog reduces its activity when it lacks food, but it continues to influence the number of prey whatever the number. The biogeocoenotic consequences of this are exceptionally important (in spite of the relatively few insects, populations of frogs numbering 1000 per ha were observed). Such observations help to explain the appearance of paradox ecosystems in which the biomass of predatory insects greatly exceeds that of phytophages, giving an erroneous impression of the intensity of energy balances within separate links of trophic chains. The potential omnivorousness of any predators (including snakes) must play a significant role in the formation of such ecosystems. This was proved by our physiological experiments more than twenty years ago (Schwarz, 1954).

The different types of adaptation of homeotherm and poikilotherm animals to their media have other biogeocoenotic consequences. To retain a body weight of some 20 g, a pair of insectivorous birds needs to consume 9 kg of food. Under the same conditions a reproducing pair of frogs consumes no less than 20 kg of insects. Similar differences exist between closely related but ecologically dissimilar species. A combination of animals, differing in the energy required to preserve their average population size, and differing in their influence on the lower trophic levels within one BGC, ensures the dynamic equilibrium of ecosystems.

Intra-and interspecific interference

A stable biogeocoenotic balance depends not only on optimal structures of BGC and separate links of food chains, nor on the number of individual species, but also on population regulation based on the specific action of animal metabolites. Studies on the larvae of amphibians and insects have shown that metabolites regulating the development and growth of animals accumulate in the media as the population density increases (Schwarz, 1972). As a result, the population becomes divided into groups of individuals developing at different rates. At any one time relatively few animals in the population grow rapidly and consume food. Influenced by the metabolites of the youngsters, these quickly complete their metamorphosis and leave the water. Vital resources are thus released for the next batch of individuals to develop. Thus food resources are used by the population with maximum effectiveness.

Interactions between organisms forming different trophic levels depend on many factors. Experiments were designed to determine the integrated effect of consumers on producer associations. Different population densities of rodents were created experimentally on confined tundra sites and on herb meadows in the South Urals, and the response of the vegetation was then observed. Detailed results of this work are published elsewhere (Smirnov & Tokmakova, 1972; Dobrynsky, 1974, in press), but the logic of the investigations and general results are given here.

The maximum production of standing crop was observed in sites where 20% of the vegetation had been consumed by *Microtus oeconomus*. This corresponds to a density of voles of about 100–200 per ha for 100 summer days, or 27–33 per ha throughout the year. As stated above, mechanisms regulating population size protect the BGC from too high densities of dominant consumers. The increase in phytomass in the study areas exceeded that of the control (Fig. 4).

The results of rodent activity were estimated by observing the growth of separate plants and changes in the structure of associations. The best index is the total photosynthetic activity of the disturbed plant community. Experiments were based on the premise that the total CO_2 consumption by all levels of the community indicates its total productivity. Different densities of *Microtus agrestis* were set up on experimental plots, and the CO_2 content of air samples was determined with an infrared gas analyser. Several days later the voles were removed and CO_2 consumption estimates were made. Subsequently regular observations on the restoration of the photosynthezising activity of the phyto-coenoses were made with regard to the density of voles in experimental plots.

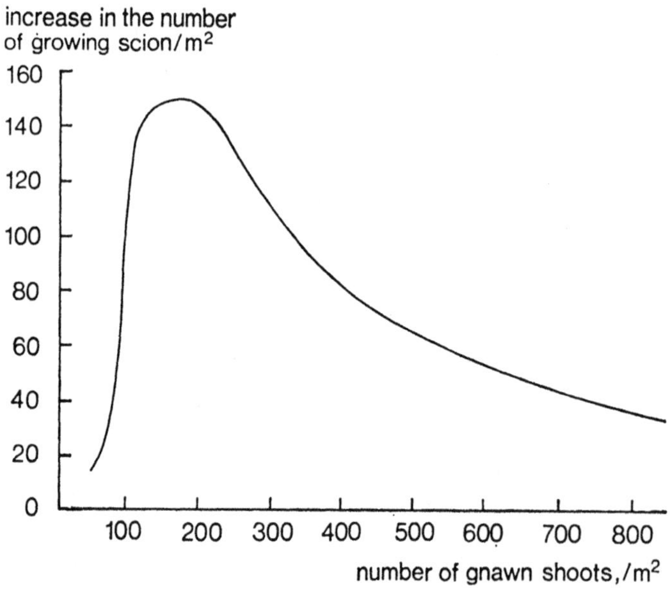

Figure 4. The response of the sedge-cotton grass-sphagnum community of producers to the influence of the consumer vole *Microtus oeconomus* in the Polar Urals, 1968–71.

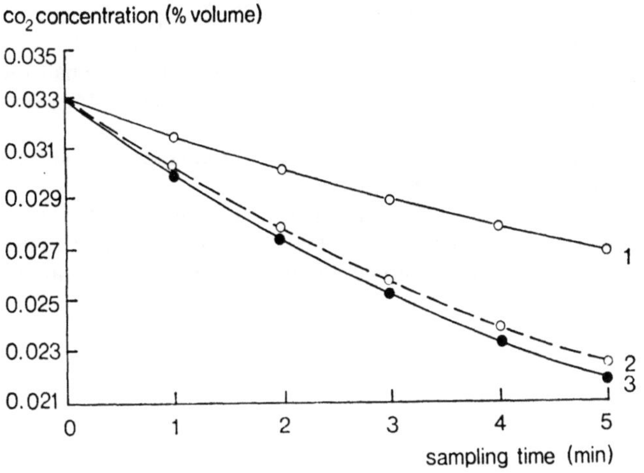

Figure 5. The ecological impact of *Microtus agrestis* voles on the total photosynthesizing activity of the phytocoenosis in 1.5 m^2 experimental plots. Curve 1 indicates the CO_2 content in air samples after one vole has been present for one week in a 1.5 m^2 plot; curve 2, the same plot a month later; curve 3, the control.

It was established that a vole population of 650/ha over one week caused some changes in the phytocoenosis which decreased the CO_2 absorption by 1.4 times. Within a month the photosynthetic activity and structure of the association was renewed (Fig. 5). But higher densities of voles (c. 1100/ha) produced such changes in the phytocoenosis that it could not be restored in one growing season.

Conclusion

The given facts all show that a harmonious development of BGC depends on population mechanisms regulating the number of dominant species, and on optimal interrelationships, fixed by evolution, between species belonging to different trophic levels.

References

Dobrynsky, L. N. 1974. Estimation of total photosynthetic activity of plant communities in different conditions. In: Materialy otchetnoj sessii zoologitscheskich laboratorij Instituta Ecologii (yearly contributions of the Zoological laboratories of the Institute of Ecology of the Academy of Sciences), Sverdlovsk, USSR (in Russian).

Isakov, J. A. and D. V. Panfilov. 1969. Zonal peculiarities of animal resources in the USSR. In: Geography of the USSR, no. 7, Institute of Scientific and Technical Information, Nauka Publishing Office, Moscow. (In Russian.)

Peshkova, N. V. 1974. Regeneration of vegetation damaged by *Dicrostonyx torquatus* during its number peak. In: Materialy otchetnoj sessii zoologitscheskich laboratorij Instituta Ecologii (yearly contributions of the Zoological laboratories of the Institute of Ecology of the Academy of Sciences), Sverdlovsk, USSR (In Russian).

Rafes, P.M. 1968. Role and significance of phytophagous insects in forest. Nauka Publishing Office, Moscow. (In Russian.)

Schwarz, S. S. 1954. An ecological analysis of morpho-physiological indices of Vertebrata. Doct. Dissertation, Moscow University (manuscript in Russian).
Schwarz, S. S. 1972. Metabolic regulation of growth and development of animals on populational and individual levels. Isvestia Akademii Nauk, USSR, no. 6, Moscow. (In Russian).
Shchupak, H. L. 1970. Dynamics of biological productivity of populations of the frog *Rana arvalis*. Ekologia, Sverdlorsk, 1 (In Russian.)
Smirnov, V. S. and S. G. Tokmakova. 1972. Influence of consumers on natural phytocoenosis production variation. In: Proceedings IV International Meeting on the biological productivity of tundra. Tundra Biome Steering Committee, Stockholm, Sweden.

Author's address:

S. S. Schwarz
Institute of Plant and
Animal Ecology
Academy of Sciences
Sverdlovsk
U.S.S.R.

Summarized by J. W. Woldendorp.

Participants: the authors E. P. Odum (U.S.A.), D. E. Reichle (U.S.A.), F. H. Rigler (Canada), H. Veldkamp (The Netherlands), together with H. Morowitz (U.S.A.), M. C. Whiteside (U.S.A.)

In the discussion several questions of methodology were posed, which concerned the falsifiability of concepts and theories. It was the general opinion that ecologists have to face the problem of forming testable statements, 'otherwise the resulting solutions would be no better than those provided by mystics' (*Whiteside*).

There appeared to be no direct questions bearing on the subject of the morning session itself, i.e. the flow of energy and matter between trophic levels. However, in the afternoon session, the communication by Professor Morowitz was followed by a discussion on the statement made by Professor Rigler concerning the cycling of energy. Although Rigler had not intended to challenge the principles of thermodynamics, part of the audience interpreted as such his thesis that there is cycling of energy in the same way as there is cycling of matter. The confusion was solved by *Morowitz* who showed that the input of the electric-magnetic energy of the sun leads to a cycling of the various elements, as is the case in all steady state thermodynamic systems. The energy of the sun is degraded into heat during this process and is lost in this way as a driving force. Therefore, there can be no closed energy cycle as occurs with the elements, of which each atom may be used anew at each point of the cycle.

Session 2
Comparative productivity in ecosystems

Chairman: F. E. Wielgolaski

Comparative productivity of ecosystems: an introduction

F. E. Wielgolaski

This subject is covered in most of the general textbooks on ecology, but the real knowledge has been limited to a few extensive studies: in particular, few studies are concerned with small animals and microorganisms. The reason for this is, of course, the tremendous and tiresome work of analysing whole ecosystems carefully, both the abiotic factors and their influence on primary, secondary and tertiary biological production. This demands real teamwork, which is often very difficult to establish because of lack of qualified manpower in the various fields, funding difficulties, and sometimes even from lack of interest from the scientists.

We need greatly increased knowledge of the productivity of various ecosystems, from tropical rainforest to tundra, and to know more about the interrelationships between the magnitudes of biological material produced in the various ecosystems. It is important to analyse the resources available to the rapidly increasing human population of the world—about 2000 million in 1930, nearly 4000 million today, a number which may be doubled before the year 2000 A.D.

In which systems and in which parts of the world can we expect to find the largest reserves of potential food for mankind? Where can we expect to find possibilities of more intensive cultivation to increase our yield of higher plants and animals, as well as of algae and other lower organisms, without destroying the balance of the various ecosystems?

Intensive productivity studies within the marine, freshwater and terrestrial sections of the International Biological Programme have increased our knowledge within various ecosystems considerably during the last 5–6 years. Many of the data, however, from the IBP field work have not yet been fully analysed, though several publications with results from these studies are now available from various parts of the world. Contributors to this section who have been very active in different biomes within the IBP will of course include as much as possible of this recent knowledge in their papers.

My work has been in the tundra biome of the IBP, and here, even under the extreme climatic conditions of alpine and polar tundra areas, a higher production of plant and lower animal life has generally been found than was estimated by earlier workers. The potential production in cool areas may be still higher, as has been shown for example by fertilizer studies, indicating that the cool and often relatively moist areas of the world may serve as a food reserve for mankind. As many of the tundra regions are important grazing areas it is of great interest to compare their production with that of the grassland biome. On the other hand there are several links between tundra and forest biomes, via for instance the forest tundra particularly in USSR and the 'Krumholz' belt in alpine tundra. Functionally the biological production of many tundra areas is comparable to the production in arid zones, although the climatic conditions, and often also the biological turnover time, are quite different in true desert areas. There are, of course, many similarities as well between the biological production of desert and grassland biomes, and between these and the forest biomes (coniferous forests and deciduous forests from cool to warm regions). The linkage between productivity of terrestial and aquatic systems is also close. This was shown for instance by the discussion within IBP on what should be included in wetland bordering both freshwater and marine biomes. Within tundra areas it is often found that the biological productivity of freshwater systems is strongly dependent on allochthonous material from the surrounding terrestrial communities. This again shows the difficulty of distinguishing between the production of various biomes and ecosystems. In some countries these problems were taken into account before IBP projects were started by combining the terrestrial and freshwater systems in an area into integrated projects.

Most productivity studies carried out both before and within IBP were however, limited to specific biomes, and often, too, to certain trophic levels. Some important comparative interbiome work was carried out earlier, particularly on primary production, by for instance Rodin and Bazilevich, Olson, and our first contributor here, Helmut Lieth. It is my hope that on the basis of, amongst others, the various IBP biome intertrophic analyses, which we hope will be published within a relatively short time, it will be possible to enlarge these to true interbiome, intertrophic comparisons. I am not sure if this can best be done by simulation and other types of modern mathematical modelling, or by traditional visual comparisons. At least may our session here be seen as the first step towards a global intertrophic comparison of biological production, of enormous importance to biological studies, and which may even be of some help in the world's food crisis. It is my hope that INTECOL will try to stimulate more work in the future on the comparative productivity of ecosystems.

Author's address:

F. E. Wielgolaski
Botanical Laboratory
University of Oslo
Blindern
Norway

Primary productivity in ecosystems: comparative analysis of global patterns

H. Lieth

Introduction

Of the pathways of energy flow through the ecosystem: sensible heat, reflection, evapotranspiration and metabolism, metabolism is usually the narrowest but nevertheless the most important road. The entrance key to this pathway is photosynthesis. The first important observable effect is plant biomass accumulation: the net primary productivity (NPP).

This NPP varies over several orders of magnitude in different parts of the globe. The variations are of vital importance for the selfmaintenance or management of the respective ecosystems, and can in most cases be attributed to environmental conditions. The understanding of these variations and their causes is therefore of prime importance for the best use of individual types of ecosystem. It is therefore my intention to present here the global productivity patterns, to consider briefly the reasons for the differences in level, and to touch on quality differences in the organic matter produced. In order to present the material in the most compact, but usable, form I shall depend largely on models and tables. Details of methods and supporting basic data may be obtained from Lieth & Whittaker (1975).

Vegetation types and productivity levels

The earth's vegetation cover consists of a generic matrix of taxa which perform the key photosynthetic processes in very similar ways but differing widely in the quantity of production and the allocation of the products. The results of these differences substantiate the various physiognomic vegetation types as we commonly discuss them in the biological sciences. Although these vegetation types are in many ways inconvenient for the discussion of global productivity patterns it is always easier to start on the basis of a known principle. In Table 1 therefore we compare the NPP of different biomes of the world. This table presents the productive power of world vegetation in two ways: dry matter accumulation and energy fixation. It is based on a table published several times (Lieth, 1972, 1973) and corrected with new data provided by Murphy (1975), Whittaker & Likens (1975), Jones & Gore (1974) and Rodin et al. (1972). Summarizing its content we find that the total terrestrial productivity amounts to 121.7×10^9 t dry

Table 1. Net primary productivity and energy fixation of major vegetation units of the world†

		NPP			Annual energy fixation			
Vegetation unit	area 10^6 km^2	range kg m^{-2} yr^{-1}	approx. mean kg m^{-2} yr^{-1}	total for area 10^9 metric tons	approx combustion value kcal g^{-1}	mean for m^2 10^6 cal m^{-2}	total for area 10^{18} cal	Authors
1	2	3	4	5	6	7	8	9
FORESTS	*50*			*81.6*			*368.6*	3
Tropical rainforest	17.0	1.0–3.5	2.8 (2.0)	47.4 (34.0)	4.1	11.5	195.5	3, 4
Raingreen forest	7.5	1.6–2.5	1.75 (1.5)	13.2 (11.3)	4.2	7.4	55.5	3, 4
Summergreen forest	7.0	0.4–2.5	1.0	7.0	4.6	4.6	32.2	3
Mediterranean sclerophyll forest (inclusive chaparral)	1.5	0.25–1.5	0.8	1.2	4.9	3.9	5.9	3
Warm temperate mixed forest	5.0	0.6–2.5	1.0	5.0	4.8	4.7	23.5	3
Boreal forest	12.0	0.3–1.2 (0.2–1.5)	0.65 (0.50)	7.8 (6.0)	4.6	3.0	36.0	3, 5
WOODLAND	7	0.2–1.0	0.6	*4.2*	4.6	2.8	*19.6*	3

† Previous estimates (2, 3) in ( ).

1	2	3	4	5	6	7	8	9
DWARF AND OPEN SCRUB	*26*			*2.6*			*11.0*	3
Tundra	8.0	0.06–1.3	0.16	1.3 (1.1)	4.5	0.7	5.6	3, 5, 6
Desert scrub	18.0	0.01–0.25	0.07	1.3	4.5	0.3	5.4	3
GRASSLAND	*24*			*19.2*			*76.8*	3
Tropical grassland (including grass-dominated savannah)	15.0	0.2–2.9 (0.2–2.0)	0.8 (0.7)	12.0 (10.5)	4.0	3.2	48.0	3, 4
Temperate grassland	9.0	0.07–1.3 (0.1–1.5)	0.8 (0.5)	7.2 (4.5)	4.0	3.2	28.8	3, 5
DESERT (extreme)	*24*			—			*0.1*	3
Dry desert	8.5	0–0.01	0.003	—	4.5	—	0.1	3
Ice desert	15.5	0–0.001	—	—	—	—	—	3
CULTIVATED LAND	*14*	0.1–4.0	0.65	*9.1*	4.1	2.7	*37.8*	3
FRESH WATER	*4*			*5.0*			*21.4*	3
Swamps and marsh	2.0	0.8–4.0	2.0	4.0	4.2	8.4	16.8	3
Lake and stream	2.0	0.1–1.5	0.5	1.0	4.5	2.3	4.6	3

Table 1. (Contd.)†

Vegetation unit	area 10^6 km^2	NPP range kg m^{-2} yr^{-1}	approx. mean kg m^{-2} yr^{-1}	total for area 10^9 metric tons	Annual energy fixation approx. combustion value kcal g^{-1}	mean for m^2 10^6 cal m^{-2}	total for area 10^{18} cal	Authors
1	2	3	4	5	6	7	8	9
TOTAL FOR CONTINENTS	149			121.7			535.3	
previous estimates in ( )				*(100.2)*			*(426.1)*	
Open ocean	332	0.002–0.4	0.13	41.5	4.9	0.6	199.2	1, 2
Upwelling zones	0.4	0.4–0.6	0.5	0.2	4.9	2.5	1.0	1, 2
Continental shelf	26.6	0.2–0.6	0.36	9.2	4.5	1.6	43.1	1, 2
Algae beds and reefs	0.6	0.5–4.0	2.0	1.2	4.5	9.0	3.6	1, 2
Estuaries	1.4	0.5–4.0	1.8	2.5	4.5	8.1	11.3	1, 2
TOTAL MARINE	*361*	0.002–4.0	0.161	*55.0*			*258.2*	2
FULL TOTAL	510			176.7			793.5	

Authors: 1. Whittaker and Likens in Lieth & Whittaker 1975*.
2. Lieth 1972.
3. Lieth 1973.
4. Murphy in Lieth & Whittaker 1975*.
5. Rodin, Bazilevich & Rozov 1972*.
6. Jones & Gore 1974*.

* Indicates paper in press.
† Previous estimates (2, 3) in ( ).

matter per year, equivalent to 2.25×10^{21} J. The marine productivity in comparison is 55×10^9 t dry matter, 1.1×10^{21} J. This new compilation gives values about 20% higher for terrestrial ecosystems. The major revision was caused by the re-evaluation of the average figures for the two tropical categories, tropical rain forest and raingreen forest.

The solar energy conversion efficiency based on the input level of 2.1×10^{24} J ($= 793.5 \times 10^{18}$ cal) annually reads with these new figures 0.16% for the total productivity of which 33% is contributed by the oceans and 67% by the terrestrial vegetation. Our previous calculations had yielded 0.13% for the global efficiency.

The increase of productivity found for tropical areas brings our calculations closer to the latest figures of Rodin et al. (1972) but our land value is still about one third smaller than theirs. It is very evident that the uncertainties still prevailing for tropical areas are the main obstacle to a better calculation. We need therefore to stress once again the urgency for more and better data sets from tropical areas (see also Brünig 1974).

The data in Table 1 may be converted into a productivity pattern map of the world. We have presented such maps in various forms (e.g. Lieth, 1964, 1972). The latest was a computer map presented at the Seattle IBP Conference in 1972 (Fig. 1). As a computer map this permits quick area-quantity calculations. This map and our previous table (Lieth, 1972) were made independently of one another but yielded similar global productivity levels (Box, 1975). The map fell short of the values in the old table by about 10% (Lieth 1972, 1973) and is about 30% short of those computed in Table 1. The reason why we have not prepared the revised 'Hague Model' map in this way will be justified in the modelling section below.

Regional productivity patterns as a base for better global patterns

A major shortcoming of Table 1 is the calculation of the mean value for each vegetation type. This was done by averaging available measurements from the respective biome types without assuring the importance or representativeness of the community chosen. It therefore seemed important to provide a more significant base for future real productivity estimates. Regional productivity maps may become far more important for future planning of land use and carrying capacity than they are now. We have undertaken such a study in North Carolina as part of the Eastern Deciduous Forest Biome (EDFB) project of the U.S. IBP. To obtain regional productivity patterns in North Carolina we used the crop statistics available and tried to assess the total net primary productivity by converting yield data into total productivity data, then multiplying the productivity data with the area covered by this crop. The averaging of the values for the major forms of land use provides a mean productivity value for the political entity listed as a subdivision in the crop yield or forestry statistics. For North Carolina it was most convenient to use the county as the basic geographic entity. North Carolina has 100 counties. With this many points it was easy to construct

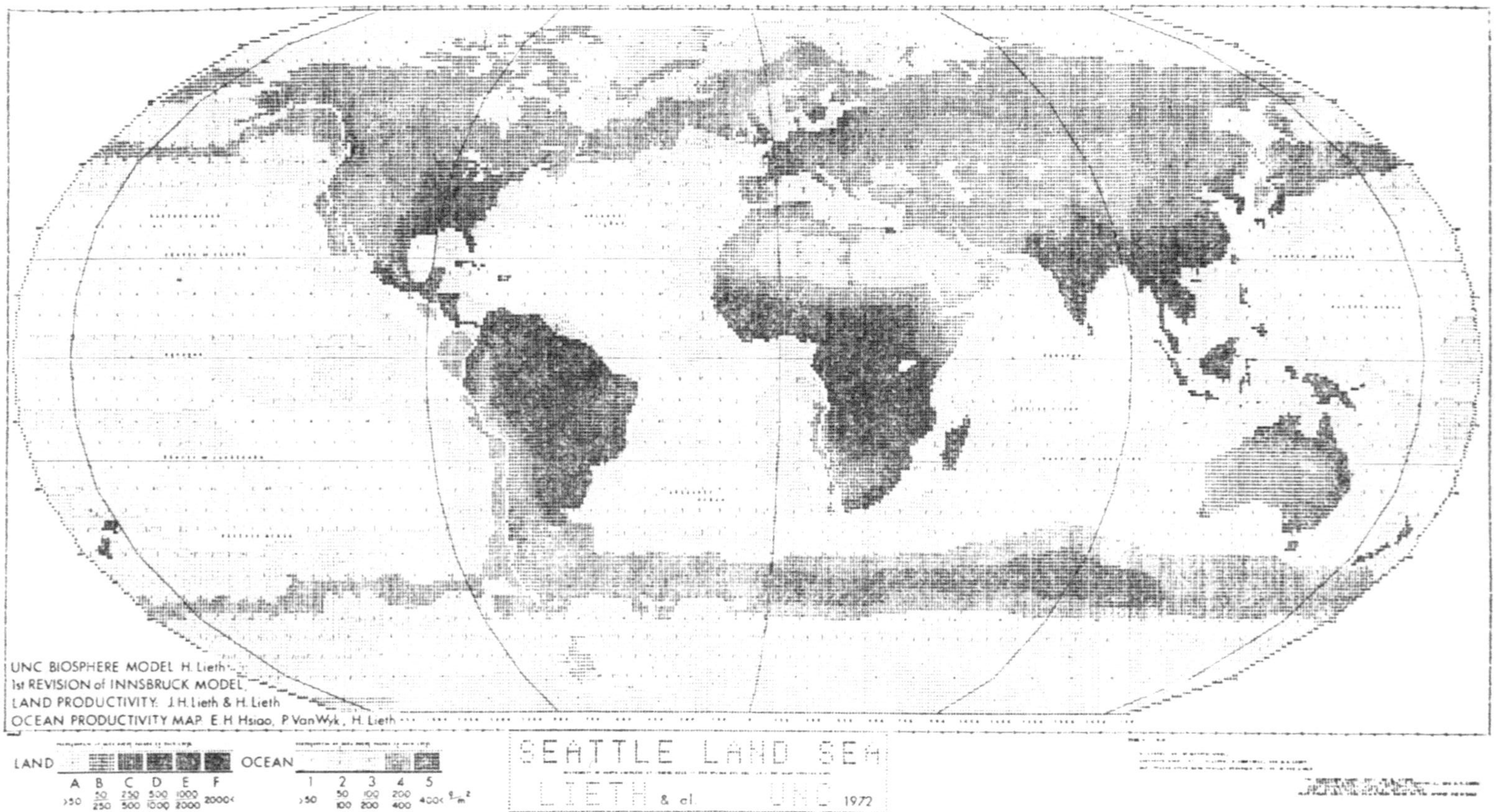

Figure 1. The Seattle productivity map based on vegetation units shown in Table 1. A computer simulation of our 1964 handdrawn map. The land portion was published as the Innsbruck productivity map in Lieth (1972). This and all other computer maps included in this paper were executed with SYMAP.

meaningful patterns for individual crops and the total productivity in the state as well. Figure 2 shows the total adjusted productivity pattern across North Carolina.

Further details including single crop production and suggestions for the improvement of the land use statistics may be obtained from Sharp et al. (1975). The final results are used here to construct models applicable on a global scale. The EDFB (Eastern Deciduous Forest Biome) of the U.S. IBP included several states wholly or partly investigated in the way described for North Carolina. The simultaneous study of Wisconsin, New York, Massachusetts, Tennessee and North Carolina now provide a north–south profile over several latitudes and altitudes. For details see Sharpe (1975), Art & Marks (1973), De Selm et al. (1971) and Stearns et al. (1971).

Together with the simultaneously run phenology projects (Hopp 1975; Reader et al. 1975) the data base was collected for a correlation study between productivity levels and length of growing season. The results are summarized in Figure 3. The average adjusted productivity values for each county are plotted in a coordinate system of length of growing period (x axis in days) and NPP (y axis in t/ha). Reader (1973) calculated for the linear correlation the equation

$$P = -1.57 + 0.0517\,S \tag{1}$$

Although constructed only for the Eastern Deciduous Forest Biome in the U.S. the model may be tested for its global usefulness by extending the equation in both directions from 0 g productivity to 365 days. If this is done we can try to use the model to convert an existing global map of the length of growing season by Wyatt & Sharp (1974) into a primary productivity map (Fig. 4).

This map is remarkable in as much as it provides us with the lowest global value that we have experienced so far with any of our models. Consistent with our past custom we will refer to this map and model combination as the 'Hague model'.

The global assessment of NPP with environmental parameters

The function of ecosystems is largely dictated by environmental forces. This is especially true for the photosynthetic process and the net primary productivity. Consequently this fact leads to the attempt to predict NPP from the ruling environmental parameters. In several recent papers (Lieth, 1973); Lieth & Box (1972) we have presented such predictions using temperature, precipitation and evapotranspiration as the environmental predictors. The combined use of average annual temperature and precipitation values for global NPP prediction was tried in the 'Miami model' (Lieth, 1973). Two equations were constructed

$$P = \frac{3000}{1 + e^{1.315-0.119T}} \tag{2}$$

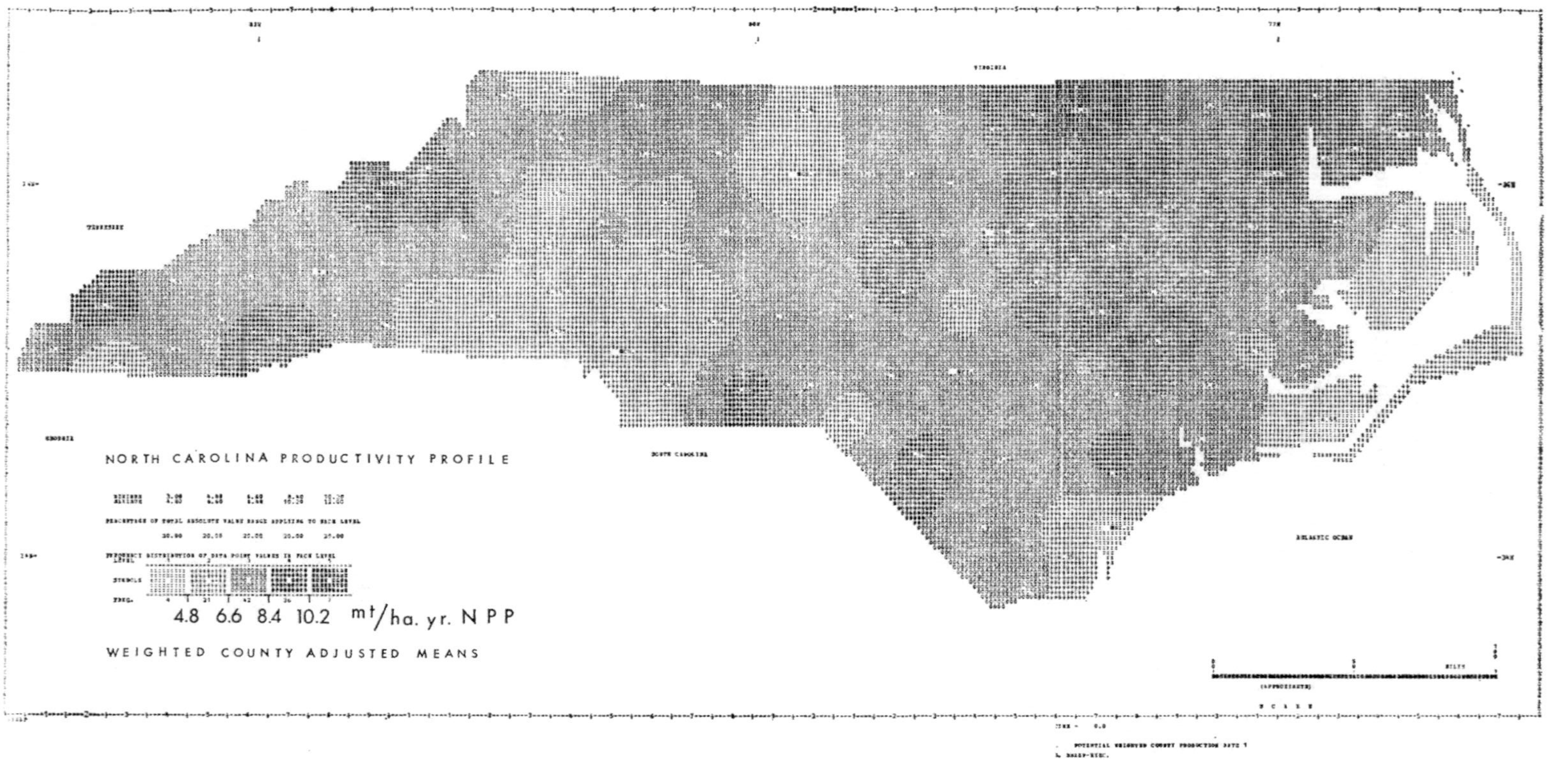

Figure 2. Productivity statistics of North Carolina. Weighted county productivity using forests, agricultural crops and other land use categories, in each case adjusting the statistical data with supporting measurements and calculations for total productivity. (These data points were used in Figure 3.)

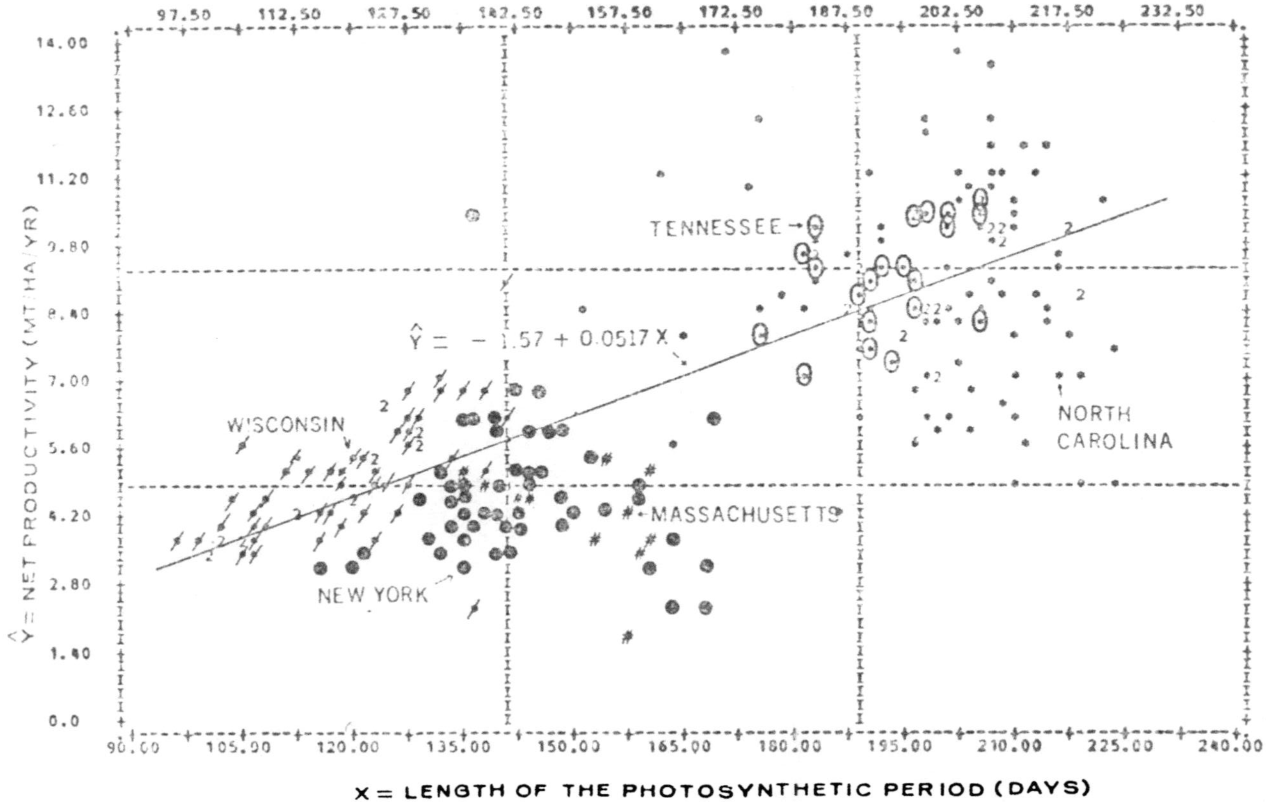

Figure 3. Net primary productivity vs. photosynthetic period. The data sets were taken from the U.S. IBP Eastern Deciduous Forest Biome memo reports for productivity profiles of Wisconsin, New York, Massachusetts, Tennessee and North Carolina.

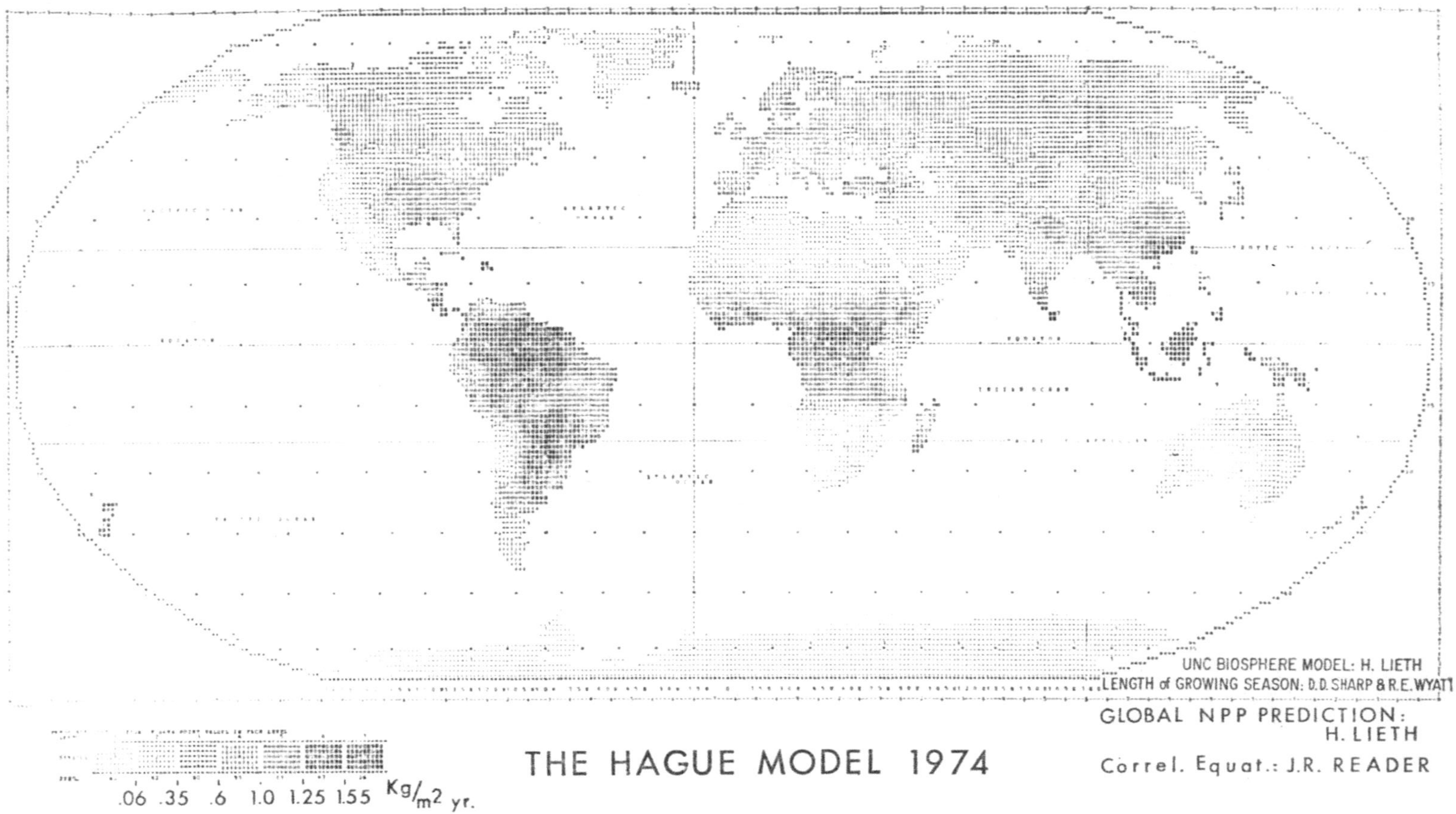

Figure 4. The Hague Model 1974: net primary productivity predicted from the length of the growing period. World growing period map based on ca. 600 points was produced by Sharp & Wyatt (1974). The predicting equation was obtained from the data set in Figure 3.

for temperature T vs. productivity, and

$$P = 3000(1 - e^{-0.000664N}) \quad (3)$$

for precipitation N, e being the natural log base.

The models were constructed using a limited but well distributed data set, (Lieth, 1972). More than 1000 stations were then selected, spread over the continents as evenly as possible, for which both temperature and precipitation data were available and the data pairs converted into NPP by selecting the lowest value if the two models yielded different levels. The NPP values were then used to prepare a computer map similar to the two already shown.

This map, (Fig. 5), known as the Miami model, was evaluated quantitatively by Box (1975). He obtained 123.9×10^9 t of dry matter for the annual terrestrial NPP, a value which almost exactly fits the respective value of our compilation in Table 1 (121.7×10^9 t).[1]

To reinforce the demonstration of the usefulness of our environmental approach to productivity assessment we have taken evapotranspiration (E in mm) to predict NPP (P in g/m^2). This exercise was presented by Lieth & Box (1972). We constructed from our data set the equation

$$P = 3000(1 - e^{-0.0009695(E-20)}) \quad (4)$$

With this model we converted the computer simulation of Geiger's world map of actual evapotranspiration into an NPP map. This map and model are referred to as 'Montreal 2: the C. W. Thornthwaite memorial model'. From this Box (1975) has calculated the global NPP value as 118.1×10^9 t per year.

Validations of models, maps and table

As new data sets are generated it will be desirable to test existing models with these. As new models are constructed it is necessary to test these against existing data sets. This procedure is most important either to reinforce or to reject existing patterns. Along with it we discuss the various opinions held on the global productivity level.

We have selected two examples pertinent to the topics so far discussed: (1) the validation of 'The Hague model' with recent data not used for the construction of the model, and (2) the testing of a new model by Terjung and coworkers (1973 and unpublished) against the terrestrial portion of the Seattle map.

A variety of models are available expressing in part the relation between NPP and the length of growing season. Also newer data sets provided the length of growing season in addition to other parameters. Several of the more recent approaches are compared in Figure 6. In this the standard line is the line R which is Reader's regression line for the Eastern Deciduous Forest Biome used to construct Fig. 4. We compare this first to the three other lines on the figure:

[1] Dr. Rodin reported in the discussion that Bodyko and Ephimova have predicted a terrestrial NPP of 134×10^9 t per year from an energy budget.

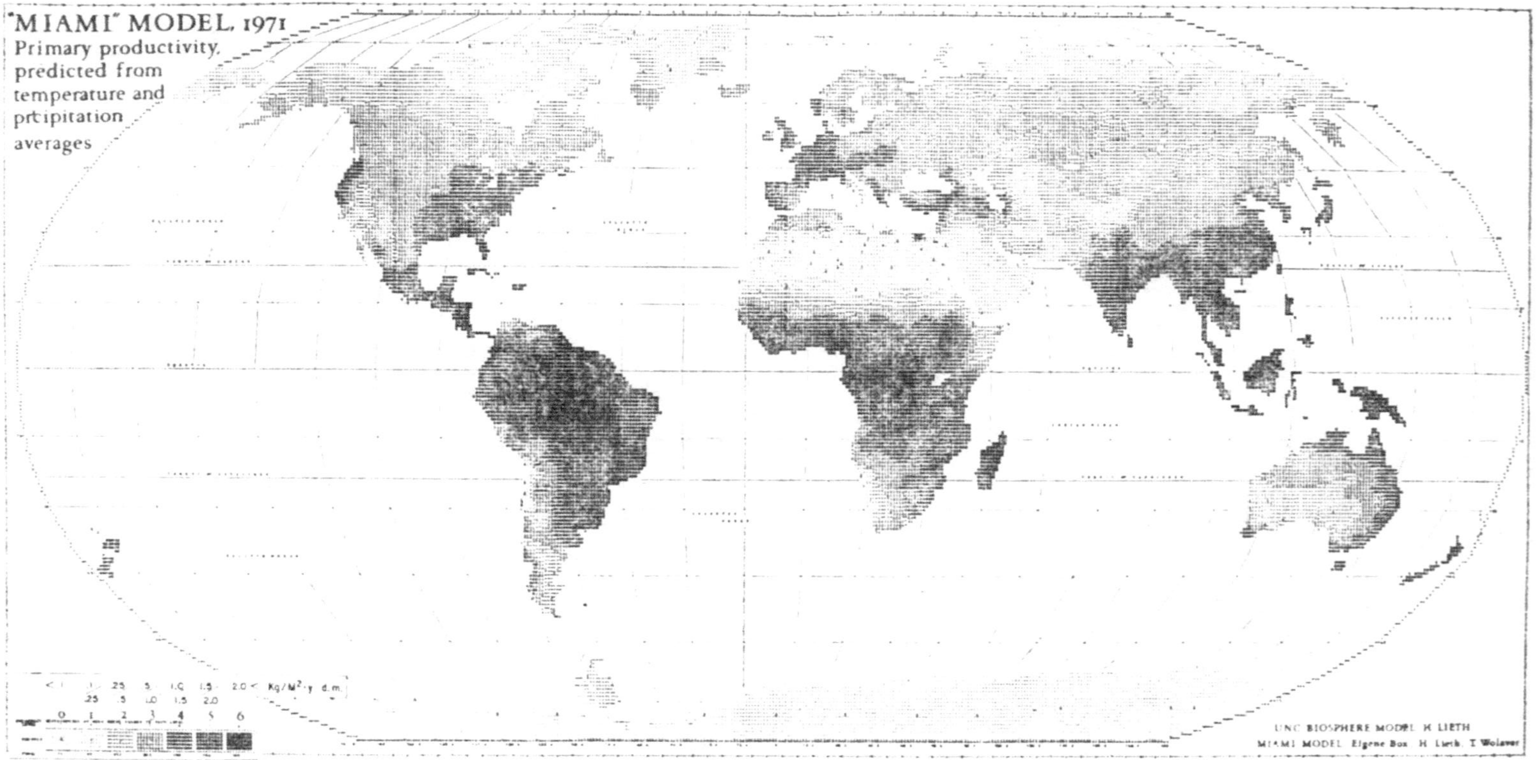

Figure 5. The Miami Model: predicting net primary productivity from precipitation and temperature. For details see Lieth (1973).

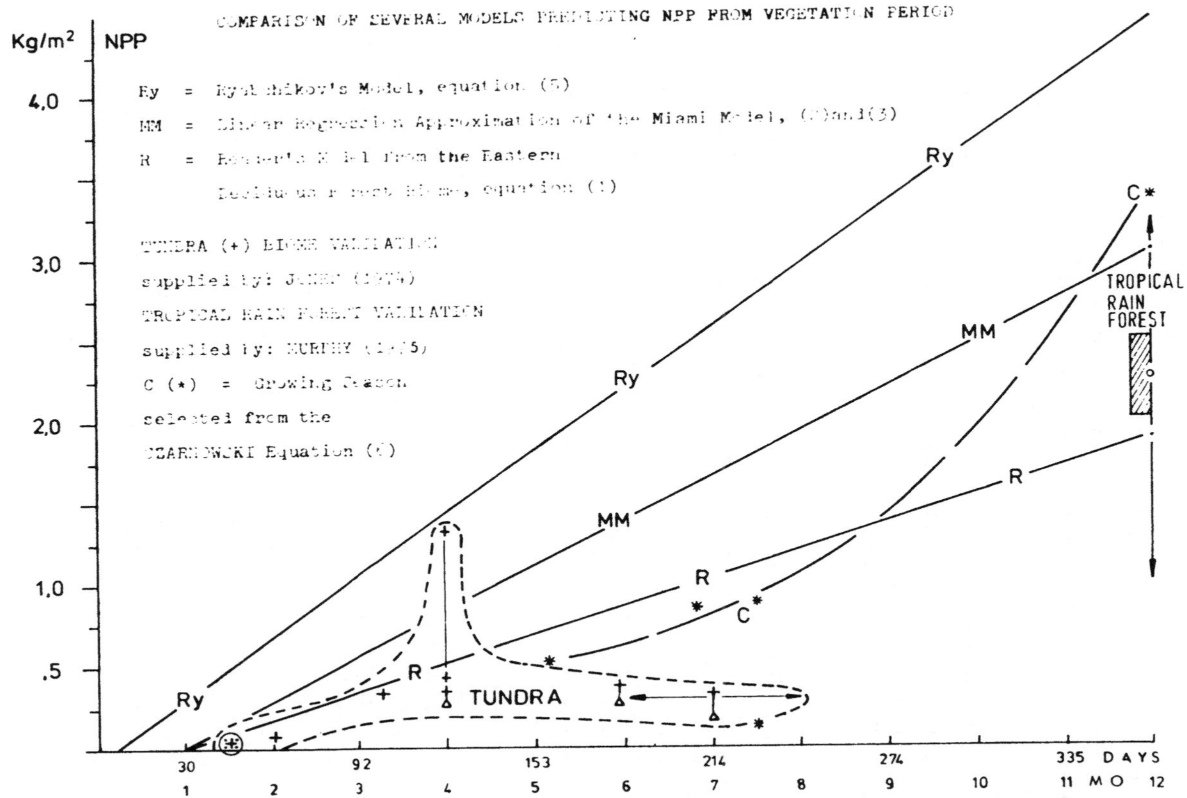

Figure 6. The comparison of several models predicting net primary productivity from the length of the vegetation period. Abscissa: vegetation period in days or months; ordinate: net primary productivity in kg/m² per year. The slopes of the lines R, MM and Ry are discussed in the text. The validation data used for tundra and tropical rain forests indicate that the global productivity level of the Miami model is probably the most accurate. This level is also supported by Table 1. For further explanation see text.

1. The line Ry which is Ryabchikov's model (1968) (from Rodin et al., 1972).

$$Kp = \frac{W\,Tv}{36\,R} \tag{5}$$

in which Kp = productive potential, W = annual effective precipitation, Tv = length of the growing season as multiples of 10 day units, and R = annual radiation. This equation aims to assess the productive potential. It is not exactly a growing season model. It may be used here as a straight line between the two fixed end points without knowing exactly what shape the curve may have between them.

2. The line MM which uses the start and end point for the Miami model to define a straight line (which may of course be curvilinear).

3. The curve drawn through the asterisks, taken from a model developed by Czarnowski (1973). In this model

$$y = cp'\,L(1 - e^{-q}) \tag{6}$$

in which L is an indication of the length of the growing period defined by temperature only, q the humidity index calculated from evaporation and precipitation, and p' the water vapor pressure for the growing season.

Comparing the four lines one can see the striking difference between Ryabchikov's line and all the others. The slopes of the lines R, MM and Ry have the approximate ratio 1:1.65:2.44. Rodin et al. (1972) give for their tabular assessment of NPP the total terrestrial value of 172×10^9 t; the Miami model as well as the data set in Table 1 of this paper (derived from the same data base) provide for a NPP of about 121×10^9 t. The ratio of the slopes for MM:Ry is 1:1.47 comparable to the ratio of the global terrestrial values derived from each data base which is 1:1.42 (Miami model vs. the compilation by Rodin et al. Using the same approach in reverse for the Hague map based on Reader's curve we should expect 73.5×10^9 t dry matter of annual terrestrial NPP to be predicted by this model.

Czarnowski's equation has not yet been used to construct a global productivity pattern. The location of the curve suggests, however, that his assessment would come out somewhere between the growth period prediction and the Miami model.

Rodin, Bazilevich & Rozov (1972) do not specifically state that they used Ryabchikov's model to complete their very detailed productivity table of various soil-determined ecosystems, but the above mentioned discussion of the slope differences between Ry and MM strongly suggests that they have used the same data base.

The two end points of Reader's equation can be validated by two data sets recently made available to us by Jones & Gore (1974) for tundra, and Murphy (1975) for the tropical rain forest. From the values we must conclude the lower portion of Reader's equation predicts rather well (it overrates slightly); whereas the upper portion of the curve underestimates the productive power of the tropical areas. The discussion of this figure demonstrates how insecure global modeling will remain until we get more solid material from the tropical areas with which to work.

Terjung and coworkers recently started a project to predict global photosynthesis through the energy budget (Terjung & Louie, 1973; Terjung et al, in press). In their still unpublished approach they have calculated several photosynthetic models. They were kind enough to supply data for this paper and in Figure 7 we show their different models of photosynthesis vs. leaf temperature. All the models shown in this figure were used by them to calculate the annual productivity for the main growing season. Their world map based on model 3 and 4 was tested against the terrestrial portion of the Seattle map (the Innsbruck map in Lieth 1973) shown earlier in Figure 1.

Figure 8 shows the results of their comparison. The stippled areas of the map are considered to be equal in both attempts. In the white areas their models overestimate whereas in the hatched areas our model appears to overestimate. The map indicates that Terjung's map based on his model 3 + 4 yields about the same global values as our Seattle model, around 91×10^9 t of dry matter per year. The tendency of the slopes as shown in the various maps is certainly verified by Terjung and coworkers, whereas the actual degree of slope remains to be refined in either our or his map. Furthermore, investigation is needed into how sensitive their world pattern assessment is to the change from model 3 to model 2, since our experience shows that agricultural landscapes are less productive than forests in warmer regions. In particular their maize model vs. the woody species model can hardly be verified by our North Carolina data from crop statistics and plantation measurements. Should their model 1, however, become validated in the future we can well expect the NPP, presently assumed to be around the level of the Seattle or Miami model, to be raised to the level of the Ryabchikov model.

The pattern of products provided by the primary producers in different biomes

The structure and function of ecosystems depends not only on the quantity but also on the quality of the dry matter produced. We assume there is no need to justify this statement but make a brief attempt here to demonstrate major differences in materials provided by the primary producers in various biomes. We restrict our first evaluation to gross categories like carbohydrates, proteins, fatty substances and ash. Mineral component is not considered here as Rodin & Bazilevich have elaborated on this matter on various occasions. (e.g. 1966). For consumers and decomposers the organic portions are usually much more important than mineral substances provided by the plants. The following Tables 2 & 3 show how we assess such quality differences and how we may compare them in the future. This study, made as a seminar exercise at the University of North Carolina, is reported in full by Lieth (1973, 1975).

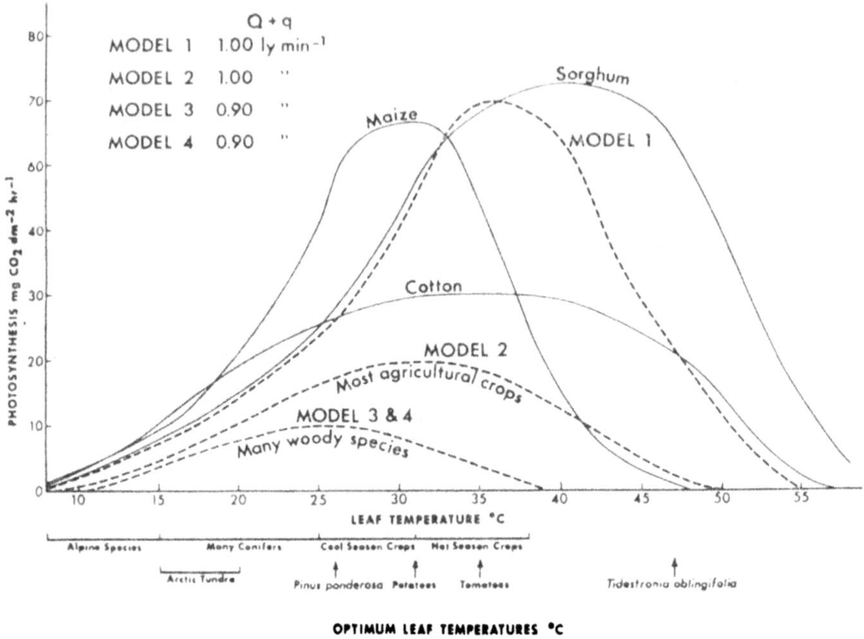

Figure 7. The various temperature-photosynthesis curves used by Terjung et al. (in press) to calculate NPP from the energy budget. Q = direct, q = indirect solar radiation, ly = cal cm^{-2}.

Table 2 shows how we assess the various qualities quantitatively. The example taken was from the grassland biome of the U.S. IBP for which we had enough details to allow a breakdown into the categories listed in the table, using the analyses provided in the Atlas of U.S. and Canadian Feeds (1971). It is clear that future studies of this kind ought to be based on our own analyses, but this approach may be justified because we first wanted to show if we can expect valuable results in terms of differences within ecosystems. The accounting procedure yields average values which we can then compare with similar computations from other ecosystems. In Table 3 we compare several ecosystem types from the temperate region. Only the grassland data are, however, assessed in the detail indicated in Table 2. The other ecosystems lack the proper data sets and need future studies in detail. The data in Table 3 are significant, however, since we do not expect changes in the figures that will overthrow the trends in the table, in which we compare a freshwater ecosystem with grassland and forest. If we analyse the trends in the separate columns we find that structural carbohydrates increase as we go from lake to forest, whereas nonstructural carbohydrates have similar levels in herbaceous communities but drop substantially in forest ecosystems. Protein has the same pattern as nonstructural carbohydrates, but deciduous forest and grassland operate with almost the same level of fat. The high

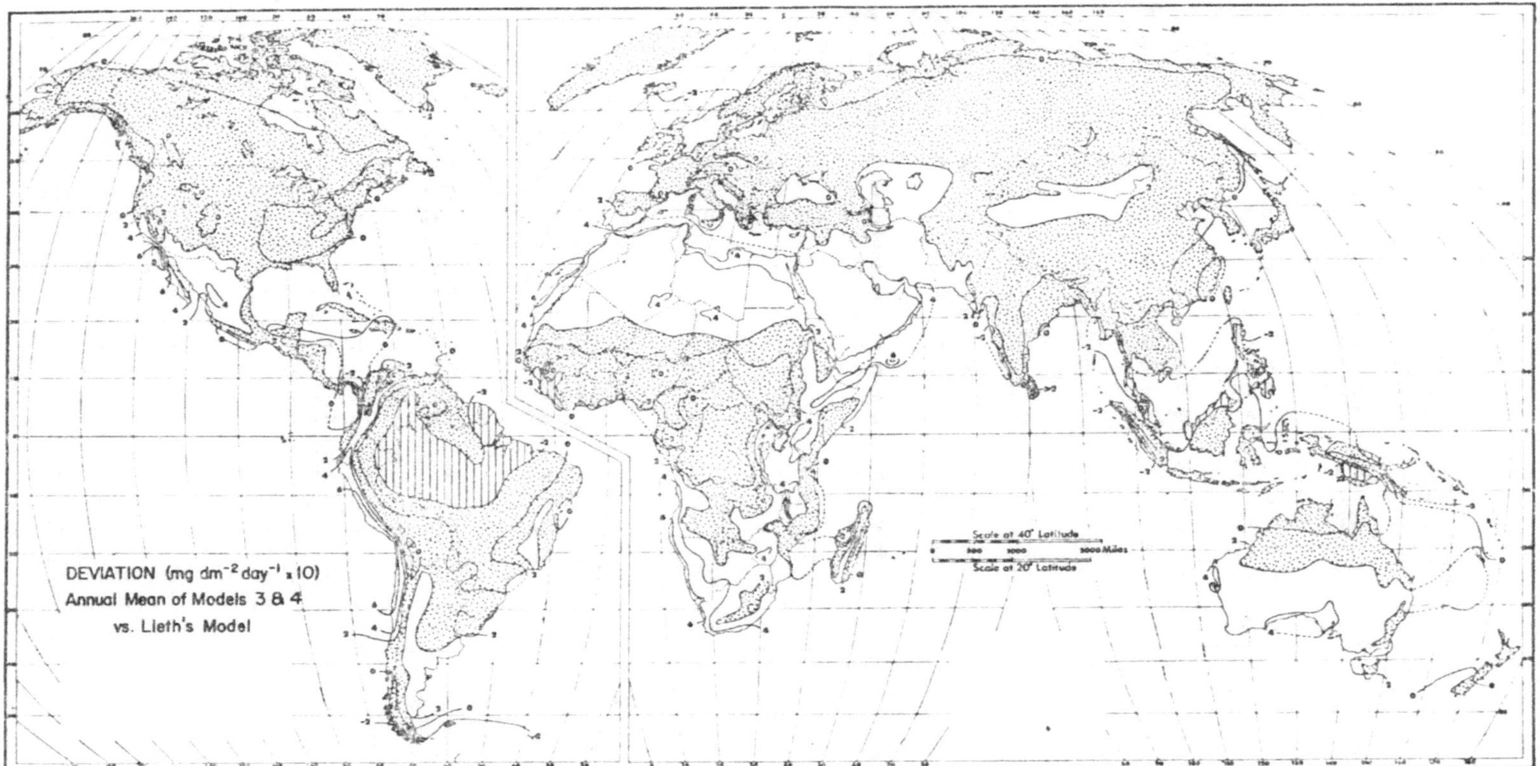

Figure 8. Testing the Terjung et al. model against the productivity map presented in Figure 1. Major deviations occur in very dry and wet areas (see text), the stippled areas are in acceptable agreement. Figures 7 and 8 are published by special permission of the authors.

Table 2. Chemical composition of above-ground grassland biomass; classes of chemicals distinguished in nutritional studies, entered separately for individual species. Table from Lieth (1975), compiled by J. R. Reader.

1	2	3	4	5	6	7	8	9	10	11	12	13	14
		Peak live biomass composition		Ash		Crude fiber		Ether extract		N-free extract		Protein	
Family or tribe	Species	%	g/m²	%	g/m²	%	g/m²	%	g/m²	%	g/m²	%	g/m²
Andropogoneae	Bluestem	24.2	384.3	6.6	25.4	34.3	131.8	2.3	8.8	50.2	192.9	6.6	25.4
Chlorideae	Grama	15.9	253.7	9.1	23.1	32.0	81.2	1.6	4.0	50.3	127.6	6.9	17.5
Compositae	Aster sp.	15.7	250.1	7.6	19.0	21.1	53.5	4.1	10.4	58.8	149.2	8.4	21.3
Festuceae	Fescue	9.5	151.7	7.5	11.4	31.9	48.4	2.9	4.4	48.1	72.9	9.6	14.6
Leguminosae	Lespedeza	6.7	107.2	7.4	7.2	44.9	48.1	2.1	2.2	32.8	35.2	12.8	13.7
Cactaceae	Opuntia	5.4	85.8	16.4	4.6	13.7	11.7	1.8	1.5	63.9	54.8	4.2	3.6
Stipae	Needlegrass	5.4	85.4	9.9	8.4	36.7	31.3	1.9	1.6	43.2	36.9	8.3	7.1
Sporoboleae	Prairie Dropseed	4.2	67.5	10.8	7.3	31.8	21.5	1.8	1.2	43.4	29.3	12.2	8.2
Triticeae	Wheat	3.9	62.4	6.5	4.0	27.9	17.4	3.8	2.4	49.8	31.1	12.0	7.5
Liliaceae	Yucca	2.3	36.9	7.1	0.5	42.7	15.7	2.0	0.7	40.5	14.9	7.7	2.8
Chenopodiaceae	Goosefoot	1.4	22.7	6.4	1.5	19.3	8.4	7.2	2.6	52.0	19.2	15.1	5.6
Paniceae	Panicum	1.4	21.4	13.1	2.8	38.3	8.2	1.3	0.3	38.8	8.3	8.5	1.8
Boraginaceae	Stickseed	0.8	12.5	11.6	1.4	10.6	1.3	3.3	0.4	51.7	6.5	22.8	2.8
Cyperaceae	Sedge	0.6	9.5	7.7	0.7	30.2	2.8	3.2	0.3	49.3	4.7	9.6	0.9
Caryophyllaceae	Silene	0.6	9.1	16.7	1.5	19.2	1.7	—	—	—	—	10.6	0.9
Danthonieae	Danthonia	0.5	7.6	4.6	0.3	30.2	2.3	3.6	0.3	52.1	3.9	9.5	0.7
Malvaceae	Mallow	0.3	5.6	16.8	0.9	21.2	1.2	2.7	0.2	36.9	2.1	22.4	1.2
Rosaceae	Avens	0.3	5·3	3·9	0.2	21.2	1.1	17.9	0.9	36.9	1.9	17.4	0.9
Aveneae	Oats	0.3	4.8	7.2	0.3	34.6	1.6	2.3	0.1	46.7	2.2	9.2	0.4
Euphorbiaceae	Euphorbia	0.3	4.3	11.0	0.4	13.0	0.5	4.1	0.2	57.5	2.5	14.4	0.6
Onagraceae	Evening Primrose	0.3	4.0	7.9	0.3	20.0	0.8	3.0	0.1	54.7	2.2	14.4	0.6
Total		100.0	1591.8	7.6	121.2	30.8	490.5	2.7	42.6	50.2	798.3	8.7	138.1

Table 3. Chemical differences in primary producers of selected ecosystem types (from Lieth 1973, 1975)
The numbers are given in % averages as extracted from literature data

Ecosystem	Comments	Ash	Structural carbohydrates* Cl.,	Hcl.,	Lgn.†	Non-structural carbohydrates	Protein	Ether extract or equivalent	Total %
Coniferous forest	Biomass, mature stand	0.3	43.5	14.5	30	1.1	1.3	7.7	98.4
	productivity	4.2	44.1	8.7	18	15.5	4.0	5.8	100.3
Summergreen deciduous forest	Standing biomass	0.3	46.6	24	20	0.8	2.5	1.8	96
	productivity	4.2	37	14.4	12	22.5	6.4	2.8	99.3
Temperate grassland	Above ground	7.6		30.8		50.2	8.7	2.7	100
	peak biomass	10		28		48	11	3	100
Freshwater lake	plankton	14		18		50	17	1.5	100
	macrophytes			14–20		43–60	8–19	1.0–2.5	
	benthic algae			9–17		36–44	5–18	0.7–2	
	Typha			30–39		38–48	7–12	1.5–3.5	
	Scirpus	6.5		33		53	7	0.5	100

* Crude fiber in table 2.
† Cl. = Cellulose, Hcl. = Hemicellulose, Lgn. = Lignin.
Table compiled as a University of North Carolina seminar project (participants H. Lieth, J. R. Reader, W. Martin, P. Carlson, G. Doyle and R. Kneib.

terpene content is largely responsible for the values in this column being twice as high in coniferous forests as in deciduous forests.

Differences as shown in Table 3 must substantially influence the types of organisms and their performance in the consumer and decomposer groups, and with further studies we expect deeper insight into the relations between decomposer and consumer species present in a given ecosystem and the proportions of chemicals offered by the primary producers. Such investigations may enable us in the future to prepare pattern maps of protein production levels, or of other organic chemical properties of the vegetation of the world. Functional geobotany that started with the global assessment of NPP is now looking forward to a new line of work that will provide us with a better insight into evaluationary processes, management priorities and future land use improvements.

Summary

Summarizing the existing information for net primary production, we find that about 177 $\times$ 10^9 t dry matter are produced by the entire biosphere equivalent to about 3.35 $\times 10^{21}$ J (793.5 $\times$ 10^{18} cal), with about 122 $\times$ 10^9 t for the terrestrial vegetation, equivalent to 2.25 $\times$ 10^{21} J (535 $\times$ 10^{18} cal). The solar energy conversion, assuming an input of 2.1 $\times$ 10^{24} J (530 $\times$ 10^{21} cal), has a mean efficiency of 0.16%, of which 67% is from terrestrial areas and 33% from the oceans. Gross patterns of global productivity from the past decade are compared. Details in Table 1.

Correlations between net primary productivity and environmental parameters can greatly facilitate regional pattern prediction. This has also been attempted on a global scale. The procedure is to build correlation models based on a limited set of actual productivity measurements paired with environmental data, construct a regression equation and then convert a large, world covering set of environmental data into biological values. So far, we have constructed four models of this type

1. $P = -1.57 + 0.0517S$ for photosynthetic season S in days (Reader).
2. $P = 3000(1 - e^{-0.000664N})$ for precipitation N in mm (Lieth, Wolaver & Box).
3. $P = (3000/1 + e^{1.315-0.119T})$ for temperature T in °C (Lieth, Wolaver & Box).
4. $P = 3000(1 - e^{-0.0009665(E-20)})$ for actual evapotranspiration E in mm (Lieth & Box). For equation (1) P reads in t/ha, for (2)–(4) in g/m^2.

Each model is convertible into a map showing the global productivity pattern. The combined use of (2) and (3) is known as the Miami Model, (4) as the Thornthwaite Memorial Model. (1) was constructed from the productivity profiles of the Eastern Deciduous Forest Biome of the U.S. It uses a data set completely different from models (2)–(4). In its present form, it reads the maximum productivity possible for 365 days as about 1800 g/m^2 whereas the other three formulae operate with a hypothetical maximum value of 3000 g/m^2. With this model we have constructed a new computer map of primary productivity based on the length of vegetation period: the Hague model. This map yields the approximate global productivity level of 73.5 × 10^9 t of terrestrial dry matter.

The primary productivity data produced in the last decade provides a starting point for future studies on a variety of scientific programs ranging from geophysics to biochemistry. From the data so far available we selected a breakdown of the organic biomass portion and compared these data for different ecosystems. Table 3 shows the differences for 3 types of ecosystem: forest, grassland and lake. Such comparisons may be used to interpret the distribution and performance of consumers and decomposers.

References

* Indicates papers and information provided before publication, used with permission of the authors.

Atlas of Nutritional data on United States and Canadian Feeds. 1971. National Academy of Science, Washington D.C.

Art, H. W. and P. L. Marks. 1973. Primary productivity profile of New York and Massachusetts. U.S.-IBP deciduous forest biome memo report 72-39. ORNL, Oak Ridge. 18 p. + tables and maps.

Box, E. 1975*. Quantitative Evaluation of the Global Productivity Models Generated by Computer. In: Primary Productivity of the Biosphere. Ecological Studies 14. H. Lieth & R. H. Whittaker, eds. Springer Verlag, New York (in press).

Brünig, E. F. 1974. Ökosysteme in den Tropen. Umschau 74: 405–410.

Czarnowski, M. S. 1973. W sprawie mapy i modelu siedliskowej zdoldosci produikcyinej Ziemi. Przeglad geograficzny 45: 295–308.

DeSelm, H. R. *et al.* 1971. Productivity profile for Tennessee. U.S.-IBP deciduous forest biome memo report 71-13. ORNL, Oak Ridge. 182 p.

Hopp, R. J. 1975. Plant Phenology Observation Networks. In: Phenology and Seasonality Modeling. Ecological Studies 8. H. Lieth, ed. Springer Verlag, New York. 25–44.

Jones, H. E. and A. J. P. Gore. 1974. Tundra Biome Synthesis Volume. Cambridge University Press, Cambridge. (in press).

Lieth, H. 1964. Versuch einer kartographischen Erfassung der Stoffproduktion der Erde. Geographisches Taschenbuch 1964/65, p. 72–80. F. Steiner Verlag, Wiesbaden.

Lieth, H. 1972. Über die Primärproduktion der Pflanzendecke der Erde. Z.f. Angew. Botanik 46: 1–37.

Lieth, H. 1973. Primary Production: Terrestrial Ecosystems. J. of Human Ecology 1: 303–332.

Lieth, H. (ed.) 1973. Chemical differences of contrasting ecosystems and their trophic levels. An exploration of a new view point in systems ecology. U.S. IBP EDF. Biome Memo Report 73-6.

Lieth, H. (ed.) 1974. Phenology and Seasonality Modeling. Springer Verlag, New York, 444 p.

Lieth, H. 1975. Some Prospects beyond Production Measurement: Comparative Analysis of some Biomass Properties on the Ecosystem Level. In: Primary Productivity of the Biosphere. Ecological Studies 14. H. Lieth & R. H. Whittaker, eds. Springer Verlag New York (in press).

Lieth, H. and E. Box. 1972. Evapotranspiration and Primary Productivity; C. W. Thornthwaite Memorial Model. In: Papers on Selected Topics in Climatology. J. R. Mather, ed. 2: 37–44, Elmer, New York.

Lieth, H. and R. H. Whittaker (eds.) 1975. Primary Productivity of the Biosphere. Ecological Studies 14. Springer Verlag, New York (in press).

Murphy, P. G. 1975. Net Primary Productivity in Tropical Terrestrial Ecosystems. In: Primary Productivity of the Biosphere. Ecological Studies 14. H. Lieth & R. H. Whittaker, eds. Springer Verlag, New York.

Reader, J. R. 1973. Phenological Investigation in Eastern North America. Thesis, Ph.D., University of North Carolina, Chapel Hill, N.C.

Reader, J. R., J. S. Radford and H. Lieth. 1975. Modeling Important Phytophenological Events in Eastern North America. In: Phenology and Seasonality Modeling. Ecological Studies 8. H. Lieth, ed. Springer Verlag, New York. 329–342.

Rodin, L. E. and N. I. Bazilevich. 1966. Production and Mineral Cycling in Terrestrial Vegetation. Oliver and Boyd, Edinburgh, 228 p.

Rodin, L. E., N. I. Bazilevich and N. N. Rozov. 1972*. Productivity of the World's Main Ecosystems. Seen as manuscript for publication in the Seattle Symposium volume, National Academy of Science, Washington, D.C. (in press).

Ryabchikov, A. M. 1968. Hydrothermal Conditions and the Productivity of Plant Mass in the Principal Landscape Zones. Vestnik MGV, Geogr. Moscow 5: 41–48, cited after Rodin, Bazilevich and Rozov 1972.

Sharp, D. D., H. Lieth and D. Whigham. 1975. Assessment of Regional Productivity in North Carolina using Cropyield Statistics. In: Primary Productivity of the Biosphere. Ecological Studies 14. H. Lieth & R. H. Whittaker, eds. Springer Verlag, New York (in press).

Sharp, D. and R. E. Wyatt. 1974. Length of growing Period in Months. Map inside back cover. In: Phenology and Seasonality Modeling. H. Lieth, ed. Springer Verlag, New York.

Sharpe, D. M. 1975. Methods for Studying the Primary Productivity of Regions. In: Primary Productivity of the Biosphere. Ecological Studies 14. H. Lieth & R. H. Whittaker, eds. Springer Verlag, New York (in press).

Stearns, F., N. Kobriger, G. Cottam and E. Howell. 1971. Productivity profile of Wisconsin. U.S.-IBP deciduous forest biome memo report 71-14 ORNL, Oak Ridge, 82 p.

Terjung, W. H. and Stella Louie. 1973. Energy Budget and Photosynthesis of Canopy Leaves. Ann. Ass. Amer. Geogr. 63: 109–130.

Terjung, W. H., Stella Louie and P. A. O'Rourke. 1974. Global Photosynthesis Model. Seen as manuscript in preparation.

Whittaker, R. H. and G. E. Likens. 1975*. Primary production: The biosphere and man. In: Primary productivity of the biosphere. Ecological Studies 14. H. Lieth & R. H. Whittaker eds. Springer Verlag, New York (in press).

Author's address:

H. Lieth
Institut für Physikalische Chemie
Kernforschungsanlage Jülich
D-517 Jülich
West Germany
and
Dept. of Botany
University of North Carolina
Chapel Hill, N.C. 27514
U.S.A.

Comparative productivity in ecosystems—secondary productivity

O. W. Heal and S. F. MacLean Jnr.

Introduction

Current emphasis upon the study of ecosystem function has generated much interest in the patterns, mechanisms and control of energy flow and productivity. The conventional approach, incorporated in the IBP Biome studies, involves collection of data on energy flow and productivity in a variety of ecosystems. These data will be compared—first in related ecosystems, and then in full variety—in search of patterns. The principles will be derived to explain the observed patterns.

In the present paper we take a different course in the comparison of secondary productivity between ecosystems. The energetic properties of consumption (C), assimilation (A), production (P), and respiration (R) of heterotrophs are derived as characteristics of the organisms' physiology and energy source. The organization of heterotrophs into trophic systems imposes further constraints upon their energetics. Thus, we examine the biological principles of secondary productivity (Wiegert & Evans, 1967; Petrusewicz & Macfadyen, 1970) and assemble them into a simple model to deduce the patterns of secondary productivity within and between a variety of terrestrial ecosystems.

This deductive approach is adopted partly because of the lack of measurements of trophic level productivity, rather than of single species productivity, in a variety of ecosystems. However the results of a few ecosystem studies, including some reported at this meeting, provide a preliminary evaluation of our deductions.

The basic carbon and energy source upon which all heterotroph production is dependent is net primary production (NPP), and this sets obvious constraints upon heterotroph production. In the balanced or steady state ecosystem total heterotroph respiration equals NPP. In an accumulating ecosystem heterotroph respiration falls below net primary productivity by some amount usually between 0 and 10% of NPP. If, then, heterotroph production is related to respiration in some simple and constant way, secondary productivity would simply track primary productivity, and the maps presented by Helmut Lieth in the previous paper could, with adjusted scale, describe secondary productivity as well. Variation in bioenergetic parameters between groups of organisms, and variation in the relative abundances of major groups of organisms in ecosystems, will lead to deviations between patterns of primary and secondary productivity.

It is a widely accepted concept that new tissue resulting from reorganization of molecules from food represents production (SCIBP, 1974). Operationally, production within a population may be determined by summing the growth increments of individuals within any period of time:

$$P = \sum_{i=1}^{n} \Delta B_i \qquad (1)$$

where ΔB_i is the change in biomass of an individual in the population within the time interval, and n is the total population size. Alternatively, production may be determined by summing change in population biomass (ΔB) and elimination (E: loss of biomass due to death, loss of body material, emigration) (Petrusewicz & Macfadyen, 1970):

$$P = \Delta B + E \qquad (2)$$

Thus, production is estimated at the population level. The sum of the productivity of heterotroph populations across a trophic sequence, e.g. herbivore-carnivore-... top carnivore, is often taken as secondary production; this represents successive reorganization of the same energy-containing molecules. Within the decomposer cycle the same principle holds, although the reorganization begins with dead organic matter. Thus, a food chain may be developed, e.g. fungi-bacteria-protozoa-bacteria, etc., on the death of the organisms. Analogous to production in the herbivore food chain, the sum of new organisms formed may be taken as secondary production.

This concept of secondary production differs from the theoretical sequence of orders of consumption and production, 1°, 2°, 3°, . . . , n° recognized in both the herbivore and saprovore trophic systems (Wiegert & Owen, 1971; Batzli, 1974). We use the term total heterotrophic productivity for the sum of successive molecular reorganizations by heterotrophic organisms and one purpose of this paper is to explore the consequences of this concept of productivity.

We use the calorie as our unit of energy flow and production, since it is a useful common denominator, but it is often used in the sense of chemically bound energy. We avoid the problem of confusing amounts e.g. production, and rates e.g. productivity, by considering all values as $m^{-2}\ yr^{-1}$.

Assimilation and growth efficiencies

A two-part taxonomic-trophic categorization of heterotrophic organisms (Table 1) summarizes a wide range of organisms and trophic functions. The taxonomic distinctions recognize a general gradient in organism size as well as the energetically important distinction between homeotherms and heterotherms. Parts of the matrix are considered blank; thus, for simplification, we treat all microorganisms as saprovores, and vertebrate microbivory is considered nil.

Table 1. A simple taxonomic-trophic categorization of heterotrophic organisms. For each category the characteristic assimilation (A/C) and growth (P/A) efficiencies are given.

	Trophic function							
	Herbivore		Carnivore		Microbivore		Saprovore	
	A/C	P/A	A/C	P/A	A/C	P/A	A/C	P/A
Microorganisms	—	—	—	—	—	—	—	0.40
Invertebrates	0.40	0.40	0.80	0.30	0.30	0.40	0.20	0.40
Vertebrate Homeotherms	0.50	0.02	0.80	0.02	—	—	—	—
Vertebrate Heterotherms	0.50	0.10	0.80	0.10	—	—	—	—

Two measures are of primary importance in the fate of ingested or consumed energy in a heterotroph population: the proportion of consumed food which is assimilated (A/C) and the proportion of assimilated food used in production (P/A). These parameters vary widely in different taxonomic and trophic groups but only a brief justification of the representative values (Table 1) can be offered here. Detailed data are referred to in Reichle (1971), McNeill & Lawton (1970) and were reviewed by Macfadyen at this Conference.

Microorganisms

The concept of consumption efficiency (A/C) used in animal ecology has little relevance for microorganisms which digest their food externally and assimilate the products of exo-digestion directly. There are also problems of comparison in terminology. For example in microbiology, assimilation often refers to the fraction of substrate which is converted to new cells whereas in fauna studies, and in the present paper, assimilation includes respiration as well as production.

The vast literature, cited by Forrest (1969), Payne (1970) and Stouthamer & Bettenhaussen (1973), on pure culture studies of bacteria, provides many values of production per unit of substrate utilized (P/A) between 0.30 and 0.60. Continuous culture techniques have tended to give lower values. Most bacterial studies have concerned growth with readily available substrates, but yield declines at low growth rate and with complex substrates. In addition there is a maintenance requirement during periods of no growth. However fungi show high values of P/A (0.25–0.50) even when growing on leaves and wood over periods up to six months (Swift, 1973, personal communication; J. C. Frankland, personal communication).

In field studies using ^{14}C labelled substrates, growth efficiencies mainly between 0.35 and 0.60 have been recorded for mixed microbial populations in soil, freshwater, estuarine and marine conditions (Shields et al., 1973; Behera & Wagner, 1974; Hobbie & Crawford, 1969; Williams, 1973). These studies used readily degradable substrates but the weight of evidence suggests that mixed populations under field conditions convert about 0.40 of the substrate utilized into cell biomass.

Invertebrates

The values of A/C for invertebrates relate to trophic function (diet) in the general pattern:

saprovores < microbivores ≈ herbivores < carnivores

(Reichle, 1971). Plant-sucking insects are often exceptions, passing large amounts of carbohydrate (energy) while feeding selectively for amino acids (Dixon, 1971); this gives them a low A/C for energy, but not for nitrogen. The general pattern is relatable to dietary quality, which might be measured as per cent long-chain polysaccharides.

The fate of assimilated energy (P/A) varies with life cycle characteristics, behaviour, growth stage, etc. (McNeill & Lawton, 1970) rather than with diet (Reichle, 1971). The highest recorded values are for larvae of parasitic insects (Pteromalidae, Ichneumonidae), with no energy demands other than those directly associated with growth (Chlodny, 1968). Lowest values are found in predatory arthropods e.g. chilopods, spiders that actively pursue their prey (Moulder & Reichle, 1972). Where energy is invested in materials not measurable as growth e.g. the 'spittle' of spittlebugs, or silk in web-spinning spiders, the production efficiency calculated from growth data is reduced.

Life cycles with long periods of maintenance of non-growing adult stages, or periods of growth cessation, lead to low P/A ratios (McNeill & Lawton, 1970). The P/A ratio is higher when calculated over the actively growing phases of the life cycle than when periods of no or negative growth are included e.g. pupae or adults indicating the growth efficiency of the organisms and of the population respectively. The range of P/A seems to be lower for terrestrial than for aquatic invertebrates (Welch, 1968).

Vertebrates

The assimilation efficiency of ungulates varies widely, with season and diet (Short et al., 1974). Grodzinski & Wunder 1975 found that assimilation efficiency of ruminants averaged around 50%, while monogastric vertebrate herbivores tended to have higher values. Non-ruminants include graminivores with an easily digested food item that is not included in the ungulate diets. On comparable diets, the ungulate mode of digestion is probably no less, and possibly more, efficient than the monogastric mode. The latter occurs in arctic Alaska, where caribou (*Rangifer tarandus*) extract a greater fraction of dry matter and energy from both monocot. and dicot. forage than do brown lemmings (*Lemmus trimucronatus*) (Batzli et al., in press).

As in invertebrates, vertebrate carnivores show high assimilation efficiency.

Carnivores feeding upon vertebrates apparently achieve higher assimilation efficiency than do carnivores upon invertebrates. This reflects the low digestibility of ingested invertebrate exoskeletons while vertebrate skeletons are often rejected prior to consumption. The arachnid and phalangid mode of ingestion results in rejection of prey exoskeleton and a high A/C, comparable to vertebrate predators upon vertebrates. Limited data on lizards (Pough, 1973) indicate assimilation efficiencies similar to homeotherms.

P/A for vertebrate homeotherms is very low compared with invertebrates, reflecting the high energy cost of thermoregulation. Maximum values generally approach 0.03 (Turner, 1970) but values for microtine rodents with high productivity and rapid population turnover may exceed 0.04 (Petrusewicz & Hansson, 1975; Batzli et al., in press). The few data on production efficiency of poikilothermic vertebrates indicate that values are intermediate between vertebrate homeotherms and invertebrates, for example, Mann (1965) quotes values between 0.07 and 0.10 for five fish species.

The values given in Table 1 and discussed above lead to some obvious predictions regarding the efficiency with which a particular population converts the energy consumed into biomass. This is the product of the two ratios, A/C and P/A, and microorganisms show high efficiency, invertebrates and heterothermic vertebrates are intermediate, and homeothermic vertebrate herbivores have very low efficiency.

Organization of trophic systems

A next level of regulation of production comes from the organization of species populations into food webs or trophic systems. Two more or less distinct trophic systems exist in virtually all ecosystems—a herbivore system based upon living autotroph tissue, and a saprovore system based on dead organic matter (Fig. 1). The two trophic systems correspond broadly to an above and below ground division, but merge to varying extents e.g. root and rhizome feeding invertebrates, which together with saprovores and microbivores, support populations of soil carnivores. Saprovorous insect larvae emerge from the soil as adults and fall prey to above ground predators.

Several important conceptual differences distinguish the two trophic systems. The herbivore system is biophagic at its base; events occurring within the system can interact with and modify the rate of input of energy into the system, as in over-grazed pastures. The saprovore system is saprophagic at its base with no direct effect upon the rate at which energy enters the system although there can be indirect effects upon primary productivity.

A second major conceptual difference lies in the pathways of energy transfer within and through the two trophic systems. In the herbivore system, the energy contained in organic molecules has two possible fates: it may be respired, with release of CO_2 and radiant energy, or it may pass to the saprovore system. The latter occurs when food is cropped but not consumed, consumed but not

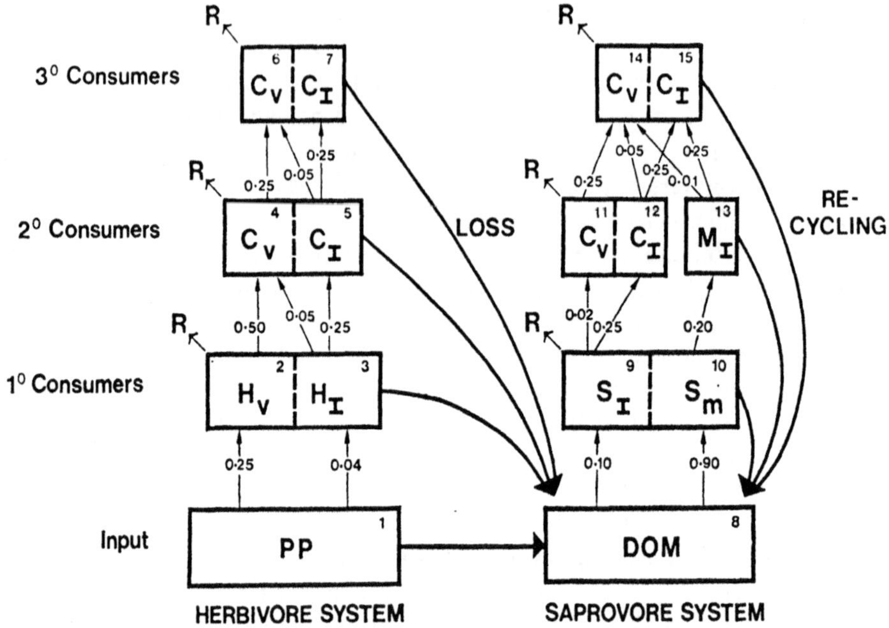

Figure 1. A generalized trophic structure for terrestrial ecosystems and the consumption efficiencies (C_n/P_{n-1}) used in the calculation of heterotrophic productivity in grasslands. PP = primary production. DOM = dead organic matter. For further abbreviations see Table 2.

assimilated, or when organisms suffer non-predatory mortality. Thus even at the first level of consumption 50% of the energy consumed passes to the saprovore system since herbivores assimilate only about 50% of the consumed food.

Once in the saprovore system, at least in a steady ecosystem, a unit of energy has but one ultimate fate: release through respiration. Energy that is 'wasted' at any trophic exchange, or energy in non-predatory mortality, returns to the pool of dead organic matter (DOM) where it is available, once more, for assimilation by microorganisms or saprovores. Thus, the saprovore system is conservative of energy through the characteristic of recycling through the pool of dead organic matter. This feature, we believe, is critical in allowing the greater length, complexity, and biomass of saprovore over herbivore food chains.

Animals of the herbivore system tend to be larger than in the saprovore system; this may reflect physical constraints on body size of soil-dwelling organisms. Vertebrates are few in the saprovore system, and function primarily at the carnivore level. Vertebrate saprovores—scavengers such as vultures—use only recently dead animals, which are only a minor part of the input to DOM, and are consequently limited in abundance in much the same way as are predators. Vertebrates are common at both the herbivore and carnivore levels of the herbivore system. Thus with the high metabolic rate of homeothermy another restriction on the flow of energy through the herbivore system is added.

Consumption efficiency

In the flow of energy through the two systems, a key parameter is C_n/P_{n-1}; consumption at any trophic level, n, to production at the previous trophic level, $n - 1$. The difference $(P_{n-1} - C_n)$ represents energy passing to the dead organic matter pool. Generalized values are given in Figure 1.

Vertebrate herbivores, especially ungulates and rodents, reach their greatest abundance in grasslands and up to 50% of available primary production may be consumed (Wiegert & Evans, 1967). This is probably an overestimate, since it is based upon above-ground plant parts. Thus, 25% of the annual net primary production is probably a more realistic maximum for terrestrial ecosystems. For most ecosystems, e.g. forests, probably less than 5% of the annual net primary production is consumed by vertebrate herbivores.

Invertebrate herbivores usually consume less than 5% of above ground primary production (Reichle & Crossley, 1967). Exceptionally, during population peaks of locusts or larvae of lepidoptera, the annual leaf production may be consumed. This still represents less than 50% of total primary production in grasslands and less than 20% in forests, although the effect on primary production may be drastic (Rafes, 1970).

Consumption of prey by carnivores is usually related to prey abundance but not prey production (Pearson, 1966, 1971). In studies of single-species population dynamics the causes of mortality can often not be quantified. The task of generalizing is further confounded by density-dependent numeric and/or functional response relationships between predator and prey populations.

In Ngorongoro in East Africa nearly all mortality of wildebeest resulted from predation by the spotted hyaena. Since the prey population was fairly stable, production equals mortality, and 100% of all production was harvested by predators. In Serengeti there were many observations of non-predatory mortality representing unharvested prey production passing to the saprovore trophic system. About 60% of the mortality of hyaenas resulted from predation (Kruuk, 1970). For *Mustela nivalis* feeding on *Microtus* in an old field, C_n/P_{n-1} was estimated at 0.37 (Golley, 1960). Thus, consumption efficiency for vertebrates preying on vertebrates may exceptionally reach 1.0, but is probably below 0.5 in most cases.

Predation by vertebrates upon invertebrates usually accounts for less than 25% of prey mortality. For example, about 20% of the mortality of spruce budworm (*Choristoneura fumiferana*) results from predation (Morris & Miller, 1954). Annual energy utilization by titmice (*Parus* spp.) and shrews (*Sorex* spp.) in an oak wood was estimated at 5% of the production of the main invertebrate prey (Varley, 1970).

Little is known of the consumption efficiency by soil fauna, especially microbivores for which it is necessary to estimate microbial production. Most estimates are indirect, but radioisotopes allow direct estimation. Coleman & McGinnis (1970) found a very small transfer from labelled fungi to soil mites

McBrayer & Reichle (1971), found that consumption by fungivores averaged 7% of dry body weight day^{-1} at field temperatures and total consumption was 24 mg fungus m^{-2} day^{-1}. Healey (1967) reported that the fungivorous collembole *Onychiurus procampatus* consumes 11 to 17% of its body weight daily. Assuming 10% and a 250-day season, annual consumption is 25 × the mean population biomass. This may account for a significant part of microbial production.

Data on consumption by invertebrate saprovores are similarly scarce. Consumption by Lumbricidae estimated from laboratory feeding rates, reaches large proportions (Satchell, 1967). In areas without Lumbricidae, consumption by litter-layer saprovores was 1% of dry body weight per day, compared with 7% for fungivores (McBrayer & Reichle 1971). Consumption by 'total saprovores' in a pine forest was 4.6% of body weight daily giving a total annual consumption of only 1% of the annual litter fall (Kowal & Crossley, 1971). Reichle (1968) found that food consumption of forest floor arthropods was related to body weight: Consumption = 0.063 $wt^{0.68}$. This gives values between those quoted above.

McBrayer & Reichle (1971) estimated consumption by soil predators in a deciduous forest to be 2.0–2.5% of body weight and 17% of the consumption by saprovores. Using the values for invertebrate saprovore A/C and P/A (Table 1) saprovore production is estimated from consumption. The resulting $C_{predator}/P_{saprovore}$ of 0.68, is high; however, the predators also feed upon microbivores, so that consumption of saprovores is less than is indicated. Based upon biomass and respiration of predators v. saprovores and microbivores, C_P/P_{S+M} is probably around 0.25.

Vertebrate carnivores may feed upon invertebrates in the saprovore system; however, their effect is small. Thus, less than 1% of the production of tipulids is consumed by major predators in moorlands (*Anthus pratensis*, meadow pipit) and in pasture (*Sturnus vulgaris*, starling) (Coulson & Whittaker, in press; Kluijver, 1933). In northern Alaska conditions are optimal for the consumption of invertebrate production by avian insectivores (MacLean, 1973) but they account for only about 10% of the production of the major prey (MacLean, in press).

Estimation of heterotrophic productivity

These values may now be applied to the general model of ecosystem organization (Fig. 1) to deduce productivity and energy flow in an ecosystem. The values for A/C and P/A (Table 1) are used for all calculations, since these are characteristics of the organisms and are not likely to change systematically between ecosystems and are based on extensive research. Values for C_n/NP_{n-1} were varied, simulating differences in abundances and ecological organization in different ecosystem types and reflecting the lack of data on consumption efficiencies. Our first calculations used a value of 0.25 which approaches maximum for consumption of net primary production by vertebrate herbivores

Table 2. Calculated ingestion, production, respiration and egestion by heterotrophs ($kcal\ m^{-2}\ yr^{-1}$) per 100 $kcal\ m^{-2}$ net annual primary production in a grassland ecosystem. Efficiencies of consumption, assimilation and production in Figure 1 and Table 1 were used in the calculation.

		Ingestion	Production	Respiration	Egestion
Herbivore system					
Herbivores,	vertebrate (H_v)	25.000	0.250	12.250	12.500
	invertebrate (H_i)	4.000	0.640	0.960	2.400
Carnivores,	vertebrate (C_v)	0.160	0.003	0.123	0.031
	invertebrate (C_i)	0.170	0.040	0.095	0.034
Saprovore system					
Saprovores,	invertebrate (S_i)	15.153	1.212	1.818	12.122
	microbial (S_m)	136.377	54.551	81.826	—
Microbivores,	invertebrate (M_i)	10.910	1.309	1.964	7.637
Carnivores,	vertebrate	0.041	0.001	0.032	0.008
	invertebrate	0.648	0.155	0.363	0.130
Total		192	58	99	35
% passing through					
Herbivore system		15.2	1.6	13.5	42.9
Saprovore system		84.8	98.4	86.5	57.1

(Fig. 1). Energy flow per 100 $kcal\ m^{-2}\ yr^{-1}$ input from net primary production was calculated from the energetic parameters in Table 1 and Figure 1, and is given in Table 2.

We reiterate that we use the concept of production in the sense of organic matter added to the biomass of a particular heterotroph species population. Through predation, biomass may be consumed, and part assimilated and either respired or added to predator biomass (production), while the remainder is lost as faeces to the pool of dead organic matter. Non-predatory mortality represents return of energy to the pool of dead organic matter from which it is, again, assimilated by saprovores; that is, it is recycled. Thus a unit of chemical potential energy, originally fixed in primary production, can be recorded in the consumption, assimilation, and production of a number of different heterotroph populations. The only constraint in the steady state (non-accumulating) ecosystem is that heterotroph respiration equals annual net primary production. Since the saprovore system is conservative of energy a unit of energy reappears in production until it is, ultimately, respired.

In the example (Table 2) 86.5% of the energy passed to the dead organic matter pool and the saprovore system: 71.0% as unconsumed primary production, 14.9% as faeces and 0.6% as non-predatory mortality of heterotrophic organisms. The herbivore system respires only 13.5% of the energy entering the system, even when herbivore consumption is set at its highest for terrestrial ecosystems. Thus, the two trophic systems of Figure 1 are clearly of two different sizes. It follows that differing levels of vertebrate herbivore ingestion, e.g. forests

(0.05) versus grasslands (0.25), have little effect upon patterns of total heterotrophic productivity.

It comes as no surprise that production and respiration of microorganisms exceeds that of animals. This high microbial production allows a production of microbivores which is greater than that of invertebrate saprovores.

Recycling within the saprovore food chain was considered by taking the energy entering the pool of dead organic matter, initially 86.6 kcal/100 kcal NPP, following this through the saprovore system, and summing egestion and non-predatory mortality to give energy re-entering dead organic matter for another pass through the system. Thus, we treat recycling conceptually as a series of discrete passes through the saprovore system.

For the system considered here, 56.7% of the energy entering DOM is lost in respiration at each pass; 43.3% is recycled in organic matter. Six passes through the saprovore system were necessary before the amount recycling fell below 1% of the NPP ($0.433^6 \times 86.6 = 0.57$). Within the decomposer subsystem theoretical distinctions of trophic levels have been made. For example, Batzli (1974) recognized 1°, 2°, . . . , *n*° detritus, representing new and recycled input into the DOM pool while saprovores (saprophages) were identified as 1°, 2°, . . . , *n*° depending upon the category of detritus used. Such distinctions are operationally impossible under field conditions, since (a) the DOM forms a spatially, structurally and chemically complex pool that is not easily subdivided, and (b) the same organisms often function simultaneously at a number of levels. Their activities cannot readily be apportioned to levels, as for example, with aboveground omnivores using a variety of discrete food types. Thus, when microbial or saprovore production is estimated from the sum of increases in population or organism size, the value includes production based upon recycling of energy containing molecules. While production cannot operationally be separated into primary and recycled components, the approach described here allows estimation of energy recycling through the DOM pool.

Evaluation of the model

Conclusions derived from such an exercise must be evaluated. We have only the most preliminary picture of total ecosystem energetics, and the lack of measurements of total net primary production, of different trophic levels and in a variety of ecosystems, reduces the possibility of evaluating adequately the accuracy of the predictions. Nevertheless, a first attempt has been made using a limited set of data for production and respiration which, to a large extent, did not contribute to the definition of the parameters used in the calculation. To calculate heterotrophic production for a site, the predicted values for each trophic level per 100 kcal m^{-2} yr^{-1} were multiplied by the measured net annual primary production for the site. For grasslands the predicted values in Table 2 were used; for non-grassland sites revised values were based on a consumption efficiency by vertebrate herbivores of 0.05.

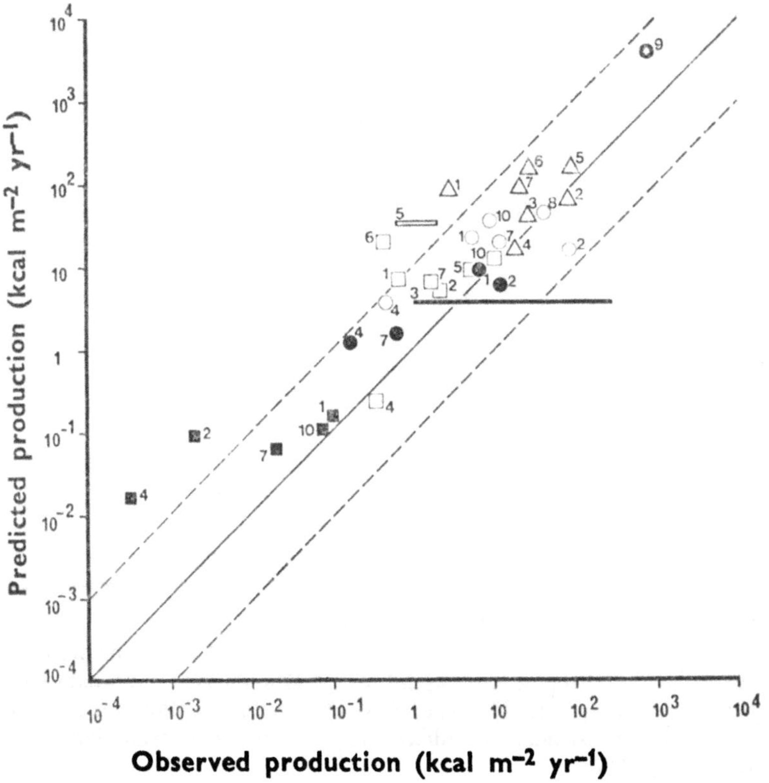

Figure 2. Heterotrophic production in a variety of ecosystems—observed and predicted values in kcal m^{-2} yr^{-1}, for vertebrate herbivores (●), invertebrate herbivores (○), vertebrate carnivores (■), invertebrate carnivores (□), invertebrate saprovores and microbivores (△), microbial saprovores (**o**). Observed values are from—Grasslands: 1, Pawnee, U.S.A. (Andrews et al., 1974). 2, Moor House, U.K. (Coulson & Whittaker, in press). 3, Point Barrow, Alaska (Brown et al., in press) including a range of annual values for vertebrate herbivores. 4, Devon Island, Canada (Bliss et al., 1973). Non-grasslands: 5, Meathop Wood, U.K. (Satchell, in press) including a range of values for invertebrate herbivores. 6, Oak Ridge, U.S.A. (Moulder & Reichle, 1972; Reichle et al., 1973). 7, Moor House, U.K. (Coulson & Whittaker, in press). 8, Virelles, Belgium (Froment et al., 1971). 9, Meerdink, Netherlands (Nagel-de Boois, 1971). 10, Wytham Wood, U.K. (Varley, 1970).

The sites used for evaluation range from tundra to tropical rain forest. The results (Fig. 2) indicate that observed values are generally within an order of magnitude of predictions. Differences between actual and predicted values may result from (a) large year to year variations some of which are indicated, (b) errors in estimation of production and respiration for field populations, as well as (c) errors in estimation made by the model.

In most cases predicted values are greater than observed values (Fig. 2). Field

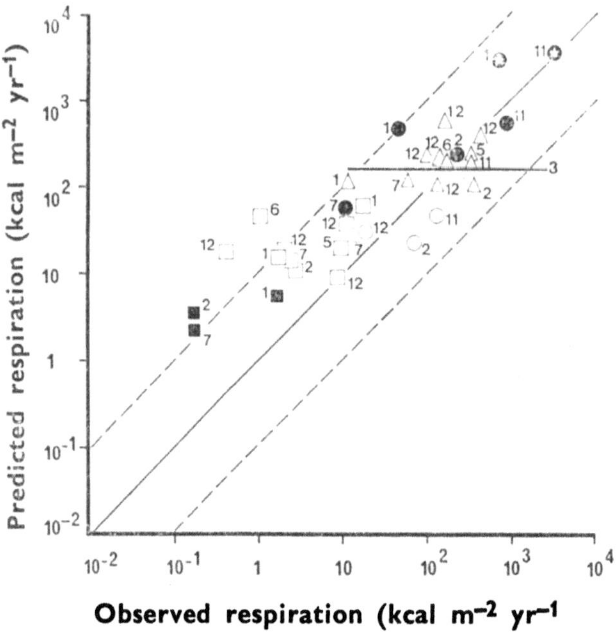

Figure 3. Heterotrophic respiration in a variety of ecosystems—observed and predicted values, in kcal m^{-2} yr^{-1}, for trophic levels as in Figure 2. Observed values are as listed in Figure 2, plus—Grasslands: 11, Armidale, Australia (K. J. Hutchinson, personal communication). Non-grasslands: 12, Alpine shrub to tropical forest, Japan and Thailand, (Kitazawa, 1970).

estimates of production rarely include all of the organisms functioning at a particular trophic level, and thus tend to be underestimates. The only observed value for annual microbial production (Nagel-de Boois, 1971) is a preliminary estimate of production of fungal mycelium. This included only the L, F and H horizons of the woodland site and was considered to be an underestimate for methodological reasons. Estimates of bacterial production per gram of soil in various temperate and tundra sites of the USSR (Parinkina, 1974) indicate annual production of the order of 10^2–10^3 kcal m^{-2} yr^{-1}.

Annual respiration predicted by the calculations is also compared with some observed values in the literature (Fig. 3). The result is similar to that for productivity. Although many of the data are for the same sites and organisms, the range of values for respiration is less than the range of production (Fig. 2). This reflects the low production but high respiratory cost of homeothermy. In both Figures 2 and 3 predicted values exceed observed values, especially in carnivores, indicating that the balance between production and respiration, determined by P/A, is reasonable. As assimilation efficiencies are based on good data and it is unlikely that site primary production is overestimated, then the errors probably result from overestimation of consumption efficiency and underestimation of observed values. Without further data, modification of model parameter values is unreasonable but it emphasises areas requiring further research. With the exception of the differences in vertebrate herbivore production in grassland and non-grassland ecosystems major variations in the pattern of heterotrophic production in different ecosystems are not recognisable, the predicted pattern of heterotrophic productivity simply follows primary productivity.

A model of this sort is a very soft one, being based upon many parameters, and variation of any one may alter the results throughout. Thus, by trial and error variation of parameters, the model output could be adjusted to fit most field data; however, this violates the deductive nature of the exercise and would provide little insight. The degrees of freedom of the model (the number of sets of input parameters that produce reasonable output) are limited by (1) the requirement that a set of output parameters, rather than one, show reasonable values simultaneously, and (2) the fact that efficiencies of assimilation (A/C) and production (P/A) are fairly well documented for most fauna and well approximated by the present values. The main areas of ignorance which affect production by soil organisms and may introduce serious errors into our predictions are (1) the proportion of dead organic matter consumed by micro-organisms and invertebrate saprovores (Fig. 2), (2) the consumption efficiency (C_n/P_{n-1}) of fauna especially microbivores, and (3) the production efficiency of mixed populations of microflora under natural conditions. Since all organisms in the saprovore subsystem are linked through the phenomenon of recycling, variation in any of these parameters has effects throughout the system. The effects may be explored with the model to indicate the range of values which produce reasonable results.

The meanings of production

Comparison of productivity within or between populations or trophic levels indicates the extent to which different organisms are able to harvest and re-organize the organic molecules of their food. Such comparisons provide insight into the functioning and ecological limitation of different populations or trophic levels, particularly in the two trophic subsystems. The apparent anomaly of heterotrophic consumption exceeding autotrophic production results from summation between trophic levels and recirculation of organic matter in the saprovore system. Indeed, there is nothing that, theoretically, prevents hetero-trophic production from exceeding primary production. Given an average P/A value of greater than 0.5, this would occur. The value of secondary production summed over a sequence of heterotroph trophic levels is not strictly comparable to the value of primary production, which represents the single trophic level. We should not compare heterotrophic assimilation with autotroph production in search of the balanced ecosystem (Andrews et al., 1974).

Because this concept of production includes recycled energy in the saprovore subsystem it is not comparable to the concept of yield. Increasing the fraction of

Table 3. Estimated annual throughput (g m^{-2}) of dry matter, nitrogen and phosphorus by microflora (M) and invertebrate (I) saprovore populations per 100 kcal m^{-2} input from primary production (NPP). Concentrations of N and P are given as % dry weight. Throughput relative to input is shown in italics.

	NPP		Microflora			Invertebrates		
	g m^{-2}	%	g m^{-2}	%	M/NPP	g m^{-2}	%	I/NPP
Dry matter	22.22	—	12.44	—	*0.56*	0.24	—	*0.01*
Nitrogen	0.33	1.5	0.75	6.0	*2.27*	0.02	10.0	*0.06*
Phosphorus	0.014	0.065	0.037	0.300	*2.64*	0.002	1.000	*0.14*

herbivore production harvested by carnivores will not necessarily influence herbivore productivity, unless the total harvest exceeds production and thus reduces herbivore biomass. In contrast, increasing the harvest of microbial production by microbivores results in increased loss of energy through microbivore and carnivore respiration, thus less energy recycling to support further microbial production.

Because of differences in efficiencies the apparent importance of different organisms varies according to the measures of consumption, production, and respiration (Table 2). For example, in these calculations, vertebrate herbivores contributed 13% of total consumption, 12% of respiration and 0.4% of production. (Clearly, the food-producing potential of vertebrate herbivores is limited.) The importance of invertebrate saprovores is highest when judged by consumption—the act by which they influence microbial activity through comminution of litter and other processes.

The accumulated production estimate for a population or trophic level, when multiplied by nutrient concentration gives the amount of nutrient cycled annually through the population. Since nutrient requirements for tissue maintenance are ignored, it is a minimum estimate of mobilization. The nutrient concentration of primary production is usually lower than that of microorganisms or fauna (Reichle, 1971); thus, nutrient turnover relative to input is greater than the relative turnover of fixed energy, or dry matter. An example, Table 3, is based on the calculated annual production of microflora and of invertebrate saprovores of 55 and 1.2 kcal per 100 kcal input from net primary production (Table 2).

The biomass turnover rate or relative production (P/B) is a useful measure of the rate of functioning of a population and is of practical value. From the ratio, based on biological theory, the productivity of a population can be estimated from easily collected biomass data (Phillipson, 1973). Annual production and biomass of microorganisms are very difficult to determine. However, accumulation of data may reveal a trend that is of predictive value. The relationship is likely to be temperature dependent and data in Parinkina (1974) and Nagel-de Boois (1971), and estimates of generation time, suggests values for P/B of 5–50.

At the opposite extreme of size and biological complexity, vertebrate homeotherms regulate their productive environment and rate of acquisition of energy. Thus the P/B ratio will vary more with size and life history strategy than with external variables such as temperature. Large, slow growing, long lived animals have a relatively low P/B, e.g. 0.2 for caribou (*Rangifer tarandus*), with seasonal breeding. Small, rapidly growing and short-lived animals have a high rate of biomass turnover e.g. about 6.0 for the brown lemming (*Lemmus trimucronatus*) with year-round breeding and a high reproductive potential (Batzli et al., in press). Phillipson (1973) suggests 2.0 and 0.2 for small and large mammals respectively. A population that is heavily harvested by predators or by man will have an age distribution favouring younger, more productive individuals and a higher P/B ratio. These characteristics correspond to "r-" and "K" life history strategies (Pianka, 1970), thus we might expect a pattern of decreasing P/B values approaching the tropics, or somewhat greater productivity per unit of vertebrate homeotherm biomass in temperate and high-latitude ecosystems. The major determinant, however, is probably the relative abundance of large (e.g. ungulate) versus small (e.g. rodent) homeotherms.

Productivity of heterotherms is strongly affected by environmental temperature, e.g. the life cycles of tundra invertebrates often extend over a number of years, in contrast to related species in more temperate areas (MacLean, 1973, 1975). Thus annual production is low. Because of the extended life cycles, a number of separate generations or year classes coexist resulting in a large abundance and biomass of individuals, each contributing a small amount to annual production; P/B is low, with decreasing values as life cycles increase. The pattern of change with latitude is opposite to that we predicted for homeotherms.

High invertebrate biomass with relatively small seasonal variation allows the development of an abundant and diverse predator fauna and a large harvest of prey production by carnivores (C_n/P_{n-1}). Thus, this parameter may change latitudinally with changing P/B relationship.

For annual species, which are more prevalent in temperate ecosystems, the population biomass replaces itself each year and P/B $\approx$ 1. In yet warmer regions several generations may be completed in one year, the biomass present at any time represents only one generation and P/B exceeds 1, increasing as the number of generations completed in a year increases.

The production to biomass ratio in invertebrates, then, may be very sensitive to temperature, length of the growing season, and other factors affecting life history characteristics. It may be very useful for predicting rate of functioning of animal populations, and possibly for microorganisms as well, from biomass data. Variations in this parameter, as well as in heterotrophic composition of different ecosystems, produce patterns of heterotrophic production which probably characterize the world's ecosystems—but which remain undefined.

This discussion has concerned the ecological organization and function of heterotrophic organisms, as formalized in a simple model. Time and space limitations have denied us the opportunity to explore the use of the model in predicting the amount and allocation among trophic types of heterotroph production in different environments or ecosystem types. Given data on abundance and rate of functioning of various taxonomic and trophic types, the patterns of productivity can be predicted.

Summary and conclusions

Currently there are few measurements of heterotrophic production for trophic levels, as opposed to species, from which patterns of variation can be examined. We have adopted an alternative approach, using results from the many single species studies on energetic efficiencies to derive a general model to predict patterns of heterotrophic production. This assumes that annual heterotrophic production is a function of

1. the input from primary production,

2. the consumption (C_n/P_{n-1}), assimilation (A/C) and growth (P/A) efficiencies of the populations which are characteristic of their taxonomic and trophic positions,

3. the organization of heterotrophs into herbivore and saprovore trophic systems.

Heterotrophic production in an ecosystem is therefore the logical consequence of the combination of these factors.

The results emphasise some fundamental distinctions between herbivore and saprovore trophic systems, in particular the conservative nature of the latter. Organic matter which is not respired by organisms within the saprovore system returns to the pool of dead organic matter where it is again available for assimilation. This recycling, in conjunction with high growth efficiency in microorganisms, results in very high levels of production within the saprovore trophic system. The estimates of accumulated production times the nutrient concentration of the organisms provides minimal estimates of nutrient throughput in the saprovore system. Nutrient incorporation in biomass is often quite high relative to nutrient input from primary production, indicating rapid nutrient recycling.

A preliminary evaluation of the model against observed values for a number of ecosystem studies shows general agreement, with heterotrophic production directly related to primary production. Predicted values tend to be higher than observed values and emphasise the need for a better understanding of consumption efficiency at all trophic levels and for a range of good observed measurements of production. These limitations prevent us from detecting variations in the pattern of production between different terrestrial ecosystems, and the present results suggest that, despite major differences in species composition, most ecosystems are similar in their organization and efficiency of heterotrophic production.

At this stage the model is very generalized and resolution is limited. It provides a means of predicting the amount of production of various taxonomic-trophic categories in ecosystems, and of exploring the effects of variations in composition and efficiencies of different groups of organisms. By adjusting the input from primary production to correspond with regional variations (Lieth, this volume) approximate estimates of heterotrophic production can be obtained on a global scale.

Acknowledgements

We are very grateful to many research workers, especially those at Merlewood Research Station and in the US Tundra Biome, for stimulating discussion and constructive criticism. The preparation of this paper was supported by the US National Science Foundation (Grant GV 29342 to the University of Alaska), under joint N.S.F. sponsorship of the International Biological Programme Tundra Biome and the Office of Polar Programs.

References

Andrews, R., D. C. Coleman, J. E. Ellis and J. S. Singh. 1974. Energy flow relationships in a short-grass prairie ecosystem. Proc. 1st Int. Congr. Ecology. Pudoc, Wageningen. 22–28.

Batzli, G. O. 1974. Production, assimilation and accumulation of organic matter in ecosystems. J. Theor. Biol. 45: 205–217.

Batzli, G. O., R. G. White, S. F. MacLean Jnr. and B. D. Collier. In press. The herbivore-based food chain. In: The structure and function of the Tundra ecosystem: synthesis volume. J. Brown, F. L. Bunnell, S. F. MacLean Jnr., P. Miller and L. L. Tieszen. eds. Dowding, Hutchinson and Ross, Philadelphia.

Behera, B. and G. H. Wagner. 1974. Microbial growth rate in glucose amended soil. Soil Sci. Soc. Amer. Proc. 38: 591–594.

Bliss, L. C., G. M. Courtin, D. L. Pattie, R. R. Riewe, D. W. A. Whitfield and P. Widden. 1973. Arctic Tundra Ecosystems. Ann. Rev. Ecol. Syst. 4: 359–399.

Brown, J., F. L. Bunnell, S. F. MacLean, P. Miller and L. L. Tieszen (Eds.) In press. The Structure and Function of the Tundra ecosystem: synthesis volume. Dowding, Hutchinson and Ross, Philadelphia.

Chlodny, J. 1968. Evaluation of some parameters of the individual energy budget of the larvae of *Pteromalus puparum* (L.) (Pteromalidae) and *Pimpla instigator* (Fabr.) (Ichneumonidae). Ekol. pol. Ser. A. 24: 505–513.

Coleman, D. C., and J. T. McGinnis. 1970. Quantification of fungus—small arthropod food chains in the soil. Oikos. 21: 134–137.

Coulson, J. C. and J. B. Whittaker. In press. Fauna of moorland soils. In: The ecology of some British moors and montane grasslands. O. W. Heal & D. F. Perkins. eds. Springer-Verlag, Berlin.

Dixon, A. G. F. 1971. Aphids. In: Methods of study in quantitative soil ecology: population, production and energy flow. J. Phillipson. ed. IBP Handbook 18. Blackwell, Oxford. 233–246.

Forrest, W. W. 1969. Energetic aspects of microbial growth. In: Microbial growth. P. Meadow & S. J. Pirt. ed. Cambridge University Press. 65–86.

Froment, A., M. Tanghe, P. Duvigneaud, A. Galoux, S. Denaeyer-de Smet, G. Schnock, J. Goulois, F. Mommaerts-Billiet and J. P. Vanseveren. 1971. La chenaie melangee calcicole de Virelles-Blaimont en haute Belgique. In: Productivity of forest ecosystems. Proc. Brussels Symp. 1969. P. Duvigneaud. ed. Unesco, Paris. 635–665.
Golley, F. B. 1960. Energy dynamics of a food chain of an old-field community. Ecol. Monogr. 30: 187–206.
Grodzinski, W. and J. Wunder. 1975. Ecological energetics of small mammals. In: Small mammals: their productivity and population dynamics. F. B. Golley, K. Petrusewicz & L. Ryszkowski. eds. Cambridge University Press. 173–204.
Healey, I. N. 1967. The energy flow through a population of soil collembola. In: Secondary productivity of terrestrial ecosystems. K. Petrusewicz. ed. Panstwowe Wydawnictwo Naukowe, Warsaw. 695–704.
Hobbie, J. E. and C. C. Crawford. 1969. Respiration corrections for bacterial uptake of dissolved organic compounds in natural waters. Limnol. Oceanogr. 14: 528–532.
Kitazawa, Y. 1971. Biological regionality of the soil fauna and its function in forest ecosystem types. In: Productivity of forest ecosystems. Proc. Brussels Symp. 1969. P. Duvigneaud. ed. Unesco, Paris. 485–514.
Kluijver, H. N. 1933. [Contribution to the biology and the ecology of the starling (*Sturnus vulgaris vulgaris* L.) during its reproductive period]. Versl. Meded. Plziektenk. Dinst Wageningen, 69.
Kowal, N. E. and D. A. Crossley. 1971. The ingestion rates of micro-arthropods in pine mor estimated from radioactive calcium. Ecology, 52: 444–452.
Kruuk, H. 1970. Interactions between populations of spotted hyaenas (*Crocuta crocuta* Erxleben) and their prey species. In: Animal populations in relation to their food resources. A. Watson. ed. Blackwell, Oxford. 359–374.
MacLean, S. F. 1973. The life cycle and growth energetics of *Pedicia hannai atenatta* Alex. (Diptera: Tipulidae), an arctic crane fly. Oikos, 24: 436–443.
MacLean, S. F. 1975. Ecological adaptations of tundra invertebrates. In: Physiological Adaptation to the Environment. J. Vernberg. ed. Intext, New York. In press.
MacLean, S. F. Jnr. In press. The saprovore-based trophic system. In: The structure and function of the Tundra ecosystem: synthesis volume. J. Brown, F. L. Bunnell, S. F. MacLean Jnr., P. Miller and L. L. Tieszen. eds. Dowding, Hutchinson and Ross, Philadelphia.
McBrayer, J. F. and D. E. Reichle. 1971. Trophic structure and feeding rates of forest soil invertebrate populations. Oikos, 22: 381–388.
McNeill, S. and J. H. Lawton. 1970. Annual production and respiration in animal populations. Nature, Lond. 225: 472–474.
Mann, K. H. 1965. Energy transformations by a population of fish in the River Thames. J. Anim. Ecol. 34: 253–276.
Morris, R. F. and C. A. Miller, 1954. The development of life tables for the spruce budworm. Can. J. Zool. 32: 283–301.
Moulder, B. C. and D. E. Reichle. 1972. Significance of spider predation in the energy dynamics of forest floor arthropod communities. Ecol. Monogr. 42: 473–498.
Nagel-de Boois, H. M. 1971. Preliminary estimate of production of fungal mycelium in forest soil layers. In: Organismes du sol et production primaire. iv. Colloquium Pedologie. Inst. Nat. Res. Agric. Paris. 447–454.
Parinkina, O. M. 1974. Bacterial production in tundra soils. In: Soil organisms and decomposition in tundra. A. J. Holding, O. W. Heal, S. F. MacLean & P. W. Flanagan eds. Tundra Biome Steering Committee, Stockholm. 65–77.
Payne, W. J. 1970. Energy yields and growth of heterotrophs. Ann. Rev. Microbiol. 24: 17–51.

Pearson, O. P. 1966. The prey of carnivores during one cycle of mouse abundance. J. Anim. Ecol. 35: 217–233.
Pearson, O. P. 1971. Additional measurements of the impact of carnivores on California voles (*Microtus californicus*). J. Mammal. 52: 41–49.
Petrusewicz, K. and L. Hansson. 1975. Biological productivity in small mammal populations. In: Small mammals: their productivity and population dynamics. F. B. Golley, K. Petrusewicz & L. Ryszkowski. eds. Cambridge University Press. 153–172.
Petrusewicz, K. and A. Macfadyen. 1970. Productivity of terrestrial animals, principles and methods. IBP Handbook 13. Blackwell, Oxford.
Phillipson, J. 1973. The biological efficiency of protein production by grazing and other land-based systems. In: The biological efficiency of protein production. J. G. W. Jones. ed. Cambridge University Press. 217–235.
Pianka, E. R. 1970. On r- and K- selection. Amer. Nat. 104: 592–597.
Pough, F. H. 1973. Lizard energetics and diet. Ecology 54: 837–844.
Rafes, P. M. 1970. Estimation of the effects of phytophagous insects on forest production. In: Analysis of temperate forest ecosystems. D. Reichle. ed. Springer-Verlag, Berlin. 100–106.
Reichle, D. E. 1968. Relation of body size to food intake, oxygen consumption and trace element metabolism in forest floor arthropods. Ecology, 49: 538–542.
Reichle, D. E. 1971. Energy and nutrient metabolism of soil and litter invertebrates. In: Productivity of forest ecosystems. Proc. Brussels Symp. 1969. P. Duvigneaud. ed. Unesco, Paris. 465–477.
Reichle, D. E. and D. A. Crossley. 1967. Investigation on heterotrophic productivity in forest insect communities. In: Secondary productivity of terrestrial ecosystems. K. Petrusewicz. ed. Panstosowe Wydawnictwo Naukowe, Warsaw. 563–587.
Reichle, D. E., R. V. O'Neill, S. V. Kaye, P. Sollins, and R. S. Booth. 1973. Systems analysis as applied to modeling ecological processes. Oikos, 24: 337–343.
Satchell, J. E. 1967. Lumbricidae. In: Soil Biology. A. Burges & F. Raw. eds. Academic Press, London, 259–318.
Satchell, J. E. (ed.) in press. The ecology of an English oakwood. Springer-Verlag, Berlin.
SCIBP (Special Committee for the International Biological Programme). 1974. Quantities, Units and Symbols. Publ. Int. Council Sci. Unions. Lond.
Shields, J. A., E. A. Paul, W. E. Lowe and D. Parkinson. 1973. Turnover of microbial tissue in soil under field conditions. Soil Biol. Biochem. 5: 753–764.
Short, H. L., R. M. Blair and C. A. Segelquist. 1974. Fiber composition and forage digestibility by small ruminants. J. Wildl. Mgmt. 38: 197–209.
Stouthamer, A. H. and C. Bettenhaussen. 1973. Utilization of energy for growth and maintenance in continuous and batch cultures of microorganisms. Biochim. Biophys. Acta. 301: 53–70.
Swift, M. J. 1973. The estimation of mycelial biomass by determination of the hexosamine content of wood tissue decayed by fungi. Soil Biol. Biochem. 5: 321–332.
Turner, F. B. 1970. The ecological efficiency of consumer populations. Ecology. 51: 741–742.
Varley, G. C. 1970. The concept of energy flow applied to a woodland community. In: Animal populations in relation to their food resources. A. Watson. ed. Blackwell, Oxford. 389–404.
Welch, H. E. 1968. Relationships between assimilation efficiencies and growth efficiencies for aquatic consumers. Ecology. 49: 755–759.
Wiegert, R. G. and F. C. Evans. 1967. Investigations of secondary productivity in

grasslands. In: Secondary productivity of terrestrial ecosystems. K. Petrusewicz. ed 2: 499–518.
Wiegert, R. G. and D. F. Owen. 1971. Trophic structure, available resources and population density in terrestrial versus aquatic ecosystems. J. Theor. Biol. 30: 69–81.
Williams, P. J. Le B. 1973. On the question of growth yields of natural heterotrophic populations. In: Modern methods in the study of microbial ecology. T. Rosswall ed. Bull. Ecol. Res. Comm. Stockholm. 17: 400–401.

Author's addresses:

O. W. Heal
Institute of Terrestrial Ecology
Merlewood Research Station
Grange-over-Sands
Cumbria
England

S. F. MacLean Jnr.
Institute of Arctic Biology
University of Alaska
Fairbanks, Alaska
U.S.A.

Energy and matter economy of ecosystems

Lech Ryszkowski

Introduction

An increased understanding of ecological structures and processes has led a growing number of ecologists to become aware of the great interpretative possibilities provided by thermodynamics or energetics. The objective of energetics is to determine the principles that govern material transformations. It is well known that various ecological structures are maintained, changed and reshaped at the expense of energy, but it is often difficult to demonstrate this fact quantitatively.

A great achievement of the International Biological Programme has been to obtain quantitative information on energy flow patterns. Much has been learnt recently about energy transformations, as well as about the factors influencing the efficiency of the flow. The stimulation provided by the IBP meant, however, that the problems of ecosystem productivity were often considered from the energetic aspect only. The influence of matter cycling on the productivity of ecosystems received much less attention, and only in the last phase of IBP were an increasing number of studies giving more time to this problem published. Until then ecological studies provided much more information on the energetic aspect of ecosystem productivity than on the cycling of matter. A description of biomass changes, even with calories as units, remains a static analysis in which the energy present in the system as a driving force is ignored. A dynamic approach, however, demands an interpretation in which energy flow and cycling of matter are combined.

Keeping in mind these well-known physical divisions of the fields of study helps us to realize what is the aim of ecological energetics at the ecosystem level. In general terms this aim is to discover the principles and laws governing matter transformations, the output of which is the ecosystem productivity. In biological systems, unlike man-made machines, the working substance (organic matter) is also the source of energy for its own transformations. In general terms the production of a system per unit time is proportional to the quantity (mass) of material available and the amount of energy for the transformation, and, because the source of raw materials is exhaustible, to its turnover rate.

Energetics of primary production

The autotrophs or primary producers of an ecosystem determine the amount of available energy. For a similar set of physical conditions the annual primary production is remarkably stable despite the different structures of the ecosystem in question (Rodin and Bazilevich, 1965; Duvigneaud, 1971; Lieth, 1972; Woodwell and Pecan, 1973). For example, in the temperate zone in ecosystems with no water deficit one can expect that the amount of energy fixed per square meter per year to be within a range of 4800 to 7200 kcal in the majority of cases, irrespective of the type of ecosystem: meadow, forest or cultivated field. This is equal to a 1–1.5% efficiency of photosynthetically active radiation. Evaluations of primary productivity are underestimated because the methods used neglect excretion and the leaching out of organic compounds from plants. Production below ground is usually poorly estimated, neglecting production of small rootlets, loss in the form of mucilage and so on. For example Samtsevich (1965) estimated that the production of the mucilage caps on the tips of wheat roots is equal to the dry weight of the grain crop from the wheat field. Practically nothing is known about the amount and variation of these components of primary production.

In terrestrial ecosystems production of algae is another unknown component of primary production. Because many soil algae are facultative heterotrophs, their trophic position is difficult to evaluate. The situation can be even more complicated because algae can be autotrophs with respect to carbon, but heterotrophs with respect to nitrogen or phosphorus (Gollerbach and Shtina, 1969). Recently Shtina (1972) made a survey of the algal biomass in various ecosystems. In spite of meagre information it seems that the biomass of algae in grasslands and cultivated fields is higher and more variable than in forests (Table 1). The soil surface blooms of algae can achieve a biomass of up to 1500 kg/ha in grasslands and up to 500 kg/ha in cultivated fields (Shtina, 1972). Few estimates of the production of soil algae exist. Using the data of Domracheva (1972) on the standing crop and productivity of Chlorophyta, Xanthophyta and Bacillariophyta obtained under seminatural conditions during one month (August), one can say that the ratio of net production to average standing crop is approximately 5. Taking this figure, and accepting that the average dry weight of algae is equal to 20% of live weight, which seems to be more realistic than estimates amounting up to 50% provided by Domracheva (1972), one can very roughly approximate the amount of energy fixed by algae assuming that their total activity is autotrophic. An average standing crop of 10 g live weight per square meter under optimal field conditions during one month can fix about 50 kcal. For the whole year the amount of fixed energy is probably less than 500 kcal/m^2. Thus, considering Shtina's (1972) data on the range of standing crops in different ecosystems (Table 1) one can assume that the amount of energy fixed by algae is less than 5% of that fixed by higher plants. Occasionally, in grassland ecosystems, this value may be expected to be higher.

Table 1. Biomass of algae in different ecosystems (after Shtina, 1972).

Forests				Grasslands				Cultivated fields		
Vegetation	Soil type	Sample depth (cm)	Biomass (kg/ha)	Vegetation	Soil type	Sample depth (cm)	Biomass (kg/ha)	Soil type	Sample depth (cm)	Biomass (kg/ha)
coniferous	humus-iron-podzol	0–10	7–20	meadow	soddy-podzolic	0–10	40–300	soddy-podzolic	ploughed layer	40–500
mesophilic oak	brown-forest	0–2	3.4–7.5	lowland meadow	bog	0–10	8–20	bog	ploughed layer	20–80
oak-larch	brown-forest	0–2	1.6–108.8	steppe herb-grass	solonetz	0–2	74			
				sedge meadow	brown forest	0–2	31.3–4155			

Heat and water balances

Various terrestrial ecosystems, forests, grasslands and cultivated fields, have different patterns of heat and water balances. Grin et al. (1970) and Rauner (1972) carried out studies on the energy and water economy in oak forest, forb steppe and barley field having very similar climatic characteristics and situated on the same type of soil—chernozem. From the amount of incoming photosynthetically active radiation (FAR) the forest intercepted the highest amount of energy, and the cultivated field the lowest amount (Table 2). The ratio of transpired water to the total rainfall also decreased in the same direction. But the efficiency of photosynthesis, calculated as the ratio of net primary production to the energy intercepted by the ecosystem, changed in the opposite direction (i.e. increased from forest to field) as also did the ratio of net production to the energetic cost of transpiration (Table 2). Thus, the forest ecosystem intercepts more energy and water, but uses them less economically than does the steppe and especially the cultivated field system. These compensating mechanisms can explain why the amount of energy fixed by primary producers is less variable among different ecosystems than are the components of heat and water balance. If the structure of the vegetation (species composition, stratification, seasonal aspects and so on) permits a greater interception of solar energy, or of inorganic compounds like water, and stabilizes microclimatic conditions, there is room for a less efficient use of the intercepted energy or water. Such a type of economy is characteristic of forest ecosystems. It should also be kept in mind that on a geographical scale the regions having a higher input of water are covered by forests.

The total amount of energy intercepted by an ecosystem can be characterized

Table 2. Photosynthetic and water energy economy of ecosystems (after Rauner, 1972).

Energetic parameters	Vegetation oak forest	forb steppe	cultivated fields [barley]
Photosynthetically active radiation [*FAR*] cal/cm^2/year (Q)	45 000	45 000	45 000
Intercepted *FAR* cal/cm^2/year (Qi)	25 000	18 000	11 000
$Qi:Q$	0.56	0.40	0.24
Precipitation [r] mm/year	750	680	680
Transpiration [E] mm/year	500	300	170
$E:r$	0.67	0.44	0.25
Net production [P] cal/cm^2/year	640	520	440
$[P:Qi] \cdot 100$	2.6	2.9	4.0
$[P:EL] \cdot 100$	2.2	2.9	4.4

L = latent heat of water vaporization [~600 cal/g].

by the value for net radiation. Heating the soil, convection of heat to the air, and the energetic cost of evapotranspiration make up the greatest contribution to the partition of the intercepted energy. The heat of the soil and the air influence the rate of chemical and biochemical reactions and thus indirectly determine productivity processes. Among terrestial ecosystems the forests, because of their complicated structure, have the greatest modifying effect on the heat distribution within the ecosystem; the rich grasslands have a moderate effect but higher than cultivated fields, while the deserts have practically none.

Volobuyev (1974), when estimating the energetic cost of soil formation, suggested that the average ratio of the energetic cost of evapotranspiration to the energy fixed by primary producers and to the energy used in weathering processes of the rock under the soil is approximately 100:1:0.01. In his calculations Volobuyev used estimates of the evapotranspiration. The transpiration index, that is the amount of transpired water per gram of organic matter produced, may be used as an approximation of the transpiration component. Taking the value of the latent heat of vaporization of water (~600 cal/g) and the calorific value of plants after Lieth (1972), the following ratios of the energetic cost of transpiration to production were calculated:

for trees (7 species) transpiration index after Assman (1968)	33.3:1
for grasses (9 species) transpiration index after Larin (1963)	76.7:1
for cultivated plants (9 species) transpiration index after Briggs and Shantz (1914)	85.0:1
and the average for all species	69.0:1

If the values obtained by Assman are correct then the trees show a remarkably efficient economy in the use of energy for transpiration.

The movement of water, the essential component of protoplasm and the carrier of nutrients, is energetically a very expensive component of production in terrestrial ecosystems. Water is therefore one of the crucial factors for productivity, as well as is the heat of the environment ('energy subsidies'—Odum, 1973). Many studies support this conclusion. (See also the review paper on the variability of primary production by Lieth in this volume.) By contrast with terrestrial ecosystems, in open water there is no transpiration by phytoplankton, and water is obtained by the energetically much cheaper processes of osmosis and diffusion. Water is also rich in carbon dioxide, and the crucial requirements for production are nutrients such as nitrogen and phosphorus.

Energy storage in ecosystems

In studies of primary production one can distinguish the following well-known patterns: (a) accumulation of organic matter in living form in forests; (b) accumulation of organic matter in soil humus in grasslands and tundra; (c) high consumption of primary production by herbivores in open waters, and (d) removal of the yield of cultivated fields by man.

Table 3. Energy accumulated in humus, plant biomass and used in weathering of rock (after Volobuyev, 1974).

Ecosystem type and soil	Energy, cal/cm² Cost of weathering (*a*) (to 3 m)	Accumulation in humus (*b*)	Accumulation in plant standing crop (*c*)	Total	$(b + c):a$
Tundra, gley	1 230	6 000	450	7 680	5.2
Taiga, podzol	2 460	6 800	14 250	23 510	8.6
Tropical rainforest, red earth	12 350	9 200	71 250	92 800	6.5
Steppe, chernozem	5 040	20 000	2 250	27 290	4.4
Dry steppe, chestnut	2 100	8 000	1 500	11 000	4.5
Semidesert, grey	3 920	4 000	750	2 670	1.2

According to estimates made by Volobuyev (1974) for terrestrial ecosystems, the highest storage of energy is in rainforest and then in steppe grassland due to the accumulation of humus (Table 3). Humus is also the greatest component of energy storage in dry grasslands and semideserts, and for quite different reasons (low temperature) in tundra. In the oceans the dissolved organic matter is the main component storing organic energy; the ratio of dissolved organic matter to living organisms is of the order of 100:1 (Riley, 1973). In freshwater ecosystems dissolved organic matter is also the greatest energy component but its ratio to living organisms is much less (Wetzel & Rich, 1973).

Energetics of plant nutrients

Rock weathering processes are very important sources of inorganic nutrients for the primary producers. The ratio of the energy accumulated in live and dead organic matter (organic energy storage) to the energy cost of rock weathering can be considered as an index of the nutrient input intensivity for plant production. Lower values of this index indicate smaller energy storing capacities in regard to energy cost of nutrient input, and therefore indicate a more intensive nutrient economy of the system in terms of energy. The lowest ratio of the energy storage in organic forms (dead + live) to the energy cost of rock weathering according to Volobuyev's (1974) estimations is found in semi-deserts and then in steppes, and the highest in forests (taiga and rainforest, Table 3). Grassland ecosystems therefore have a more intensive nutrient economy than forests and (because of low temperature) tundra.

The more intensive nutrient economy of grassland ecosystems is also reflected by a higher ash content in plant biomass. According to Rodin & Bazilevich (1965) one kg of ash is contained in 20 kg of plant biomass in deserts and in 30 kg of plant biomass in steppes, whereas in deciduous and subtropical forests one kg of ash is contained in 35 kg of plant biomass and in coniferous forests in 50–100 kg of plant biomass.

Thus a low ratio of the energy accumulated in live and dead organic matter to the energy cost of rock weathering coincides with a high ash content in the plant biomass of the ecosystem.

Energetics of secondary production

The patterns of the use of primary production that have just been discussed result from the prevailing physical conditions (the heat and water regimes) as well as from the activity of heterotrophs. Due to the stimulation of the IBP a vast amount of information has been gathered concerning the energetic characteristics of the metabolism of heterotrophs, as well as on energy flow, trophic efficiency, impact on vegetation etc. (e.g. Burges & Raw, 1967; Byzova, 1972; Duvigneaud, 1971; Phillipson, 1970, 1971; Kajak and Hillbricht-Ilkowska, 1972; Odum, 1971; Palmen, 1971; Petrusewicz, 1967; Petrusewicz and Macfadyen, 1970; Petrusewicz & Ryszkowski, 1969; Reichle, 1970; Winberg, 1968). Only the general results from these numerous studies will be discussed here.

Heterotrophs influence the turnover of matter in ecosystems by their metabolic processes and by behavioural activities such as burrowing, destruction of plants and so on. The contribution of heterotrophs to the process of mineralization of organic matter in ecosystems can be measured by their respiration. The well-known relationship between body weight and respiration describes the mineralization efficiency of heterotrophs of various sizes. Reichle (1971) estimated parameters of the allometric equation $y = ax^b$ (where y is respiration in $\mu l\ O_2 hr^{-1}$ and x is live weight in mg) for 107 values from four phyla, Nematoda, Mollusca, Annelida and Arthropoda, to be equal to $y = 0.339x^{0.808}$. For the sake of comparison the allometric equation was calculated for the same four phyla of invertebrates using the data gathered by Byzova (1972). All data were adjusted to 15°C using Reichle's $Q_{10} = 2$ approximation. The calculated equation is $y = 0.357x^{0.813}$ for 199 used values (Fig. 1). The similarity between these two equations is almost perfect. One can safely assume, therefore, that in general the respiration of litter and soil meso- and macrofauna is proportional to the 0.808th power of the body weight. The mammals have a lower level of metabolism than birds (parameter '*a*' in Table 4). Passerine birds have a higher level of metabolism than non-passerine birds. Of the anurans only the Bufonidae have a statistically significantly lower metabolic rate (Hutchison et al., 1968, Table 4). The higher value of '*b*' for a given group of animals indicates the greater influence of differentiation of body size classes on mineralization. Thus the change in size structure in populations of Nematoda or Collembola can

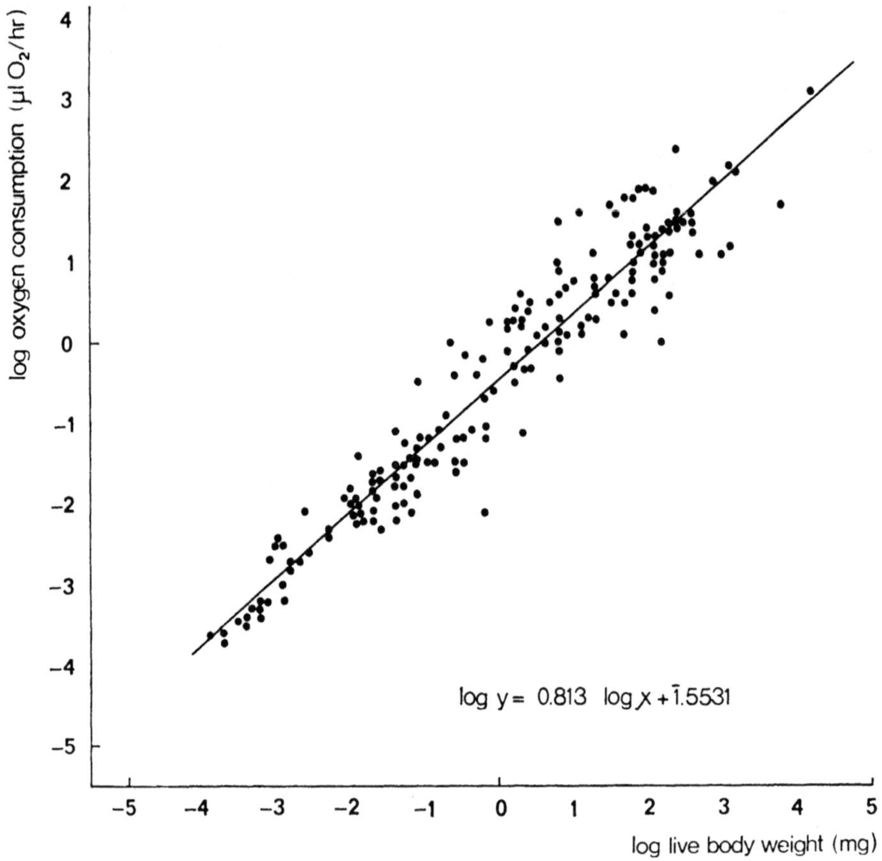

Figure 1. Live bodyweight—respiration relationship at 15 °C ($N = 199$).

Table 4. Body size and metabolism in terrestrial vertebrates (allometric equation $y = ax^b$).

Group	Units of measurement y	x	a	b	Authority
mammals	kcal/day	kg	70	0.75	Kleiber (1961)
birds	kcal/day	kg	86.4	0.66	Lasiewski & Dawson (1967)
passerine spp.			129	0.72	Lasiewski & Dawson (1967)
non-passerine spp.			78.3	0.72	Lasiewski & Dawson (1967)
anurans (15°C):					
Bufonidae	cm^3 O_2/hr	g	0.471	0.59	Hutchison et al. (1968)
Hylidae	cm^3 O_2/hr	g	0.123	0.82	Hutchison et al. (1968)
Ranidae	cm^3 O_2/hr	g	0.244	0.64	Hutchison et al. (1968)
Pelobatidae	cm^3 O_2/hr	g	0.072	0.94	Hutchison et al. (1968)

Table 5. Relationship between respiration and live body weight in various soil invertebrates described by the parameters of equation $y = ax^b$ (y = respiration $O_2 \mu l/g/hr^{-1}$, x = live weight in g). All estimations for 20 °C. (After Byzova, 1972).

Group	Range of body weight	a	b
Nematoda	$56 \cdot 10^{-6}$–$1 \cdot 10^{-6}$	117.7	0.84
Annelida			
Lumbricus terrestris	4.64–0.56	84	0.71
Dendrobaena octaedra	0.36–0.1	119	0.71
Lumbricus castaneus	0.42–0.06	86	0.64
Eiseniella tetraedra	0.14–0.06	62	0.63
Eisenia foetida	0.58–0.09	80	0.86
Eisenia rosea	0.59–0.08	88	0.95
Lumbricus rubellus	1.10–0.39	93	0.85
Allolobophora caliginosa	1.3–0.45	78	0.91
Octolasium lacteum	1.5–0.65	72	0.58
Oniscoidea	0.8–$9 \cdot 10^{-4}$	61.3	0.60
Myriapoda			
Schizophyllum caspidum	0.6–0.1	28	0.72
Japonaria laminata	0.5–0.1	70	0.82
Collembola	$8 \cdot 10^{-3}$–$7 \cdot 10^{-6}$	123.6	0.84
Scarabaeidae (larvae)			
Amphimallon solstitalis	0.93–0.013	213	0.82
Cetonia aurata	2.7–0.3	128	0.55
Melolontha hippocastani	2.5–0.07	171	1.01
Diptera (larvae)			
Tipula peliostigma	0.27-0.03	115	0.52
Oribatei	$6 \cdot 10^{-4}$–$1 \cdot 10^{-6}$	23	0.64

influence organic matter mineralization to a greater extent than can changes in size structure of Oniscoidea or Oribatei (Table 5). These are general characteristics, but one has to keep in mind that differences between species of the same systematic group are enormous, for example the range of '*b*' between earthworm species is from 0.58 to 0.95. The same differences can be observed for larvae of the Scarabaeidae (Table 5). So far it is hard to distinguish a clear pattern of differentiation of '*b*' values distribution among various groups of invertebrates. The distribution of size classes versus the abundance of animals in each class seems to be of the allometric equation type (Ghilarov, 1967), but the lack of data prevents an estimate of the patterns of influence of the size structure on the mineralization processes in different ecosystems. Nevertheless, a review of information on the sizes of invertebrates scattered throughout the literature indicates that the average weight of some groups, such as the Lumbricidae, the Collembola, many families of Coleoptera, and the Aranea, is greater in forests than in meadow or especially in cultivated field ecosystems. At least for these groups one can expect a higher impact on the mineralization of organic matter in grasslands and cultivated fields.

The secondary production is a measure of the capacity to store organic matter by heterotrophs. McNeill and Lawton (1970) analysed the relationship between population respiration and production of aquatic and terrestrial poikilotherms and land homoiotherms and distinguished the following patterns. The homoiotherms store organic matter rather inefficiently; the ratio of production to assimilation (respiration + production) is between 1.4 and 1.8%. For poikilotherms they tentatively distinguished three types: (a) short-lived poikilotherms with low respiratory cost resting stages but having a high annual population production efficiency; (b) short-lived ones with high respiratory cost non-productive stages but having intermediate population production efficiencies, and (c) long-lived poikilotherms with high cost non-productive periods and low population production efficiency.

A functional grading of invertebrates as saprovores, herbivores, and predators shows respectively increasing efficiencies of production (Reichle, 1971). The range of variation within each group is high, reflecting various adapations of the animals. The saprovores in general can be characterized as the group having the lowest production efficiencies, and therefore their role in mineralization processes is greater than, for example, herbivores.

McNeill and Lawton's classification can be applied to soil bacteria and fungi. Many authors have found a great discrepancy between evaluation of metabolic activity of microbes and the available organic matter using the parameters of production efficiencies obtained under laboratory conditions (for example Clark & Paul, 1970). The laboratory estimates of microbial production efficiencies are low, though recently high values (about 62%) were reported (Wiebe, 1971) both for aerobic and anaerobic bacteria when new methods of evaluation were used.

Thus, assuming that Wiebe's estimates apply to oligotrophic bacteria (for example Arthrobacter-like forms) which constitute the majority of soil forms one can speculate that soil bacteria under field conditions may in fact belong to the class of poikilotherms having high production efficiencies. If this is true, then often calculated discrepancies between metabolic activity of bacteria and resources of organic matter in soil should be re-examined.

Among small mammals two patterns of population production efficiencies have been recognised (Grodzinski & French, 1974). Analyses of estimates of the production and respiration of 41 populations indicate that the mean production efficiency of rodents is 2.28% and for six populations of shrews is 0.66%. Both rodents and shrews show very low production efficiencies in comparison to invertebrates. This is due to the high respiratory cost of homoiothermy in small mammals. Frenchel (1974) analysed the relationship between body size and the intrinsic rate of population increase '*r*' for a range of weights from 3.3×10^{-16} g (T-phage) to 6.1×10^{5} g (cow) including 44 estimates for microbes, aquatic and terrestrial poikilotherms, and homoiotherms. Again the allometric equation fits the data, and the separate equations for microbes, poikilotherms and homoiotherms were estimated by Frenchel (1974). If '*r*' values

Table 6. Respiration and potential production in different types of organisms of comparable size (after Frenchel, 1974).

	Unicellular organisms		Poikilotherm metazoa	Poikilotherm metazoa		Homoiotherm metazoa
ratio of respiration	1	:	8.3	1	:	28
intrinsic rate of increase	1	:	2	1	:	1.7

can be interpreted as a potential measure of production per unit weight of a population, then unicellular organisms have higher relative storing capacities (retention) for organic matter than do the metazoa (Table 6). If the estimates discussed above are correct, then one can expect both respiration and efficiency of production to increase with decreasing body size. A tentative explanation which can help one to grasp this surprising situation is the evaluation of the energy cost of maintaining the increasingly complex structure of the metazoa.

Interactions between autotrophs and heterotrophs

Many studies published recently within the framework of the IBP deal with direct pathways of energy flow between primary producers and heterotrophs at the population or groups of populations level (see the synthesis attempts cited above). The direct influence of herbivores on plants (consumption) in land ecosystems is small, except in outbreak situations, and probably only rarely reaches 10%. Thus, for example, in 18 case studies the direct impact of small mammals on their preferred vegetation was on average about 3‰, including outbreaks in which consumption amounted to 22% of the biomass of the preferred plant species (Golley et al., 1974). In comparison with the total primary production these figures run much below 0.1% of it. Such studies focus on energy flow, food webs etc. of subunits of ecosystems. There are numerous studies concerning direct transfer of energy within ecosystems. The feedback of heterotrophic activity on primary producers is less well known. These activities include destruction of plants, with an impact on the number of progeny surviving to the next growing season, transfer of plant nutrients by burrowing, earth-moving activities and so on. These impacts can be grouped into four main categories (Golley et al., 1974):

1. those concerned with destruction of a component,
2. those concerned with movement of materials,
3. those concerned with alteration of the environment,
4. those concerned with other heterotrophs.

All these types of feedback can influence the productivity processes of an ecosystem. The evaluation of their importance in different types of ecosystems is not at present possible, as the information available is too meagre, but for

Table 7. Mineralization of plant debris in various types of land ecosystems (after Kovda, 1971).

	Half-bog soil forests	Low bush tundra	Taiga	Temperate deciduous forests	Sub-tropical forests	Rain-forests	Steppe	Savannas	Temperate cultivated fields*
Time of mineralization in years	>50	20–50	10–17	3–4	~0.7	~0.1	1–1.5	~0.2	~1

* Ryszkowski unpublished data.

the sake of illustration some examples are cited briefly here. It is well known, but rarely estimated quantitatively, that herbivores destroy much more plant material than the amount needed for metabolic requirements; for instance the rodent *Microtus arvalis* in alfalfa plantations destroys twice as much as it actually eats (Ryszkowski et al., 1973). The influence of heterotrophs on the breakdown of litter, woody materials and roots was stressed a long time ago. That the animals speed up the decay of organic matter has been shown by many studies (see recent reviews by Edwards et al., 1970; Kurcheva, 1971; Zlotin & Hodashova, 1974). Nevertheless, there are studies which fail to show any impact by animals (meso- or macrofauna) on the mineralization of organic matter, especially in ecosystems intensively managed by man (for example Ryszkowski, 1974a). The overall effect on the mineralization of organic matter of the physical, chemical and biological components can be approximated by the ratio of the yearly fall of plant debris to the average standing crop of litter. According to this index in the temperate and subtropical zone grasslands and cultivated fields show a more intensive mineralization than do forests (Table 7). In grasslands, where practically all the above-ground primary production dies down quickly and is shed when the growing season is over, the decomposition processes are of great importance because the accumulation of organic matter in litter can depress primary production by changing physical conditions or by locking up nutrients in the inactive biomass (Golley et al., 1974). The opinion has been expressed that grassland humus is formed by root debris only because the above-ground vegetation is completely mineralized by animals and fires (Kucera & Kirkham, 1971). To a small extent the same phenomena have been reported for forest ecosystems. An outbreak of the herbivorous *Tortrix viridana* can induce a higher primary production of trees in the following year due to a more rapid release of nutrients tied up in the plant biomass (Zlotin & Hodashova, 1974). Thus outbreaks of herbivores or saprovores under these conditions function as part of the mechanisms that control the mineral cycling of the ecosystem. In this context, the meaning of control or stability at the population level may be quite different from that at the ecosystem level of organization (Golley et al., 1974).

The influence of animals on primary production by transferring, dislodging etc., materials across the soil profile is very poorly understood. Plant debris transferred into the soil may influence such diverse factors as soil water storage and soil temperature. The results of burrowing activity are even more striking, since the chemicals dislodged can be an important source of nutrients as well as controlling soil pH, especially in some dry grassland ecosystems. It should be kept in mind that the effects of burrowing activity are much more energetically efficient for the transfer of chemicals than is their passage through the body of animals during metabolic processes.

Biomass and species composition of heterotrophs in several ecosystems

There have been few attempts to estimate the total abundance of heterotrophs in various ecosystems. Recent syntheses have been made for example by Kovda (1971) and Whittaker and Likens (1973). In spite of many reservations these provide an overall picture of the distribution of heterotrophs in various types of ecosystems. Thus the forest ecosystems have the highest animal biomass, especially the rainforest. The aquatic ecosystems, except for upwelling zones and estuaries, have animal biomasses lower than those in savannas and temperate grasslands. Thus the retention of organic matter in the animal component of an ecosystem is highest in forests, which may be connected with the larger sizes of forest animals as discussed above. The distribution of microbes is not correlated with the amount of soil humus (Kononova, 1968); the highest abundances of microbes per gram of soil were found in brown and grey soils, while chernozem soils have only a moderate abundance of microbes. In contrast to the microbes, the abundance of soil macrofauna was well correlated with the amount of humus (Fig. 2). All these general surveys of heterotrophs should be considered as very rough approximations because turnover rates were not included in estimates, to say nothing about the accuracy of standing crop evaluations.

Comparisons of more detailed site studies of a forest ecosystem (Satchell, 1971) with cultivated field (Ryszkowski, 1974) disclose the following differences in the contributions of animals to the mineralization processes. Both projects are classified as a total ecosystem study concerned with energy flow and matter cycling. Meathop wood is a mixed deciduous forest situated in the English Lake District. The soil is a brown earth with a mull humus (Satchell, 1971). The studies on cultivated fields were carried out in the neighbourhood of Turew, Western Poland. The field soil is here a lessive type with low organic matter content; the crop rotation has been simplified to the use of potatoes, rye, and rye for aftercrops. (Ryszkowski, 1974). Due to these aftercrops the above- and below-ground primary production of the cultivated field exceeds the value achieved by the Meathop Woods (Table 8). Despite the fact that the total respiration of soil decomposers was similar in both ecosystems, the contribution of soil invertebrates to mineralization processes was about four times less in the cultivated field than in the forest (Table 8). Such agriculture reduces the abundance of soil invertebrates, resulting in their lower metabolic activity. With the greater size of animals in question one can expect the greater impact of an agricultural activity. For example, the respiration of earthworms was almost tenfold higher in Meathop Woods than in Turew cultivated fields. Groups like

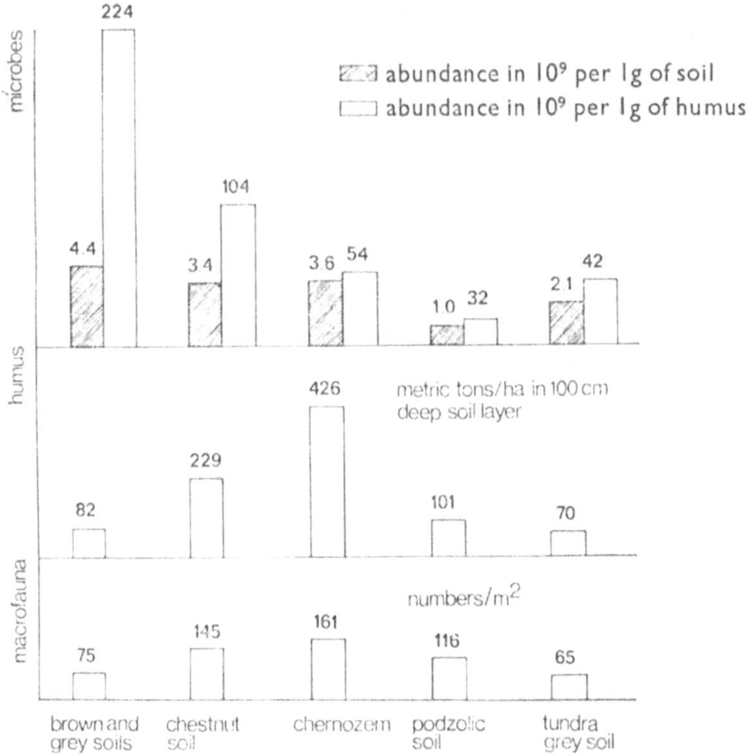

Figure 2. Distribution of humus, abundance of microbes and macrofauna in various soil types (after Kononova, 1968; Kurcheva, 1971).

Mollusca, Isopoda, and Myriapoda found in Meathop Woods were absent in cultivated fields. Dabrowska-Prot et al. (1974) who analysed the distribution of Diptera in various ecosystems showed that increased interference by man in the ecosystem generally leads to a percentage increase of the Diptera in the total entomofauna, besides a simultaneous decrease in their total number, as well as a simplification of species structure and an increase in herbivorous forms. Similar large effects of agricultural management can be observed for other groups of invertebrates.

The great similarity of the estimates for microbial respiration obtained in both studies is very questionable. In the Turew cultivated field the total respiration of soil heterotrophs was obtained from the energy balance of all inputs and outputs in the ecosystem (Golebiowska et al., 1974). Then the share of invertebrate respiration estimated separately was subtracted from total decomposer respiration to obtain the microbial respiration. In the Meathop forest study the estimates of biomass (plate counts) and the respiration rates obtained for laboratory

Table 8. Energy flow in forest and cultivated fields ecosystems ($kcal/m^2/year$).

	Cultivated fields* rye-potatoes	Meathop Wood† oak, ash, birch, sycamore
Photosynthetically active radiation (FAR)	452 000	327 807
Net primary production (Pp)	3 367–7 059	6 233
(Pp:FAR) · 100	0.7–1.5	1.91
Input of plant debris into soil	1 445–4 696	3 560
Respiration of herbivores	2–14	7
Total respiration of decomposers	1 940–2 180	2 172
Protozoa	?	12
Nematoda	30–50	85
Lumbricidae	5–6	54
Enchytraeidae	26–30	167
Other invertebrates	8	43
Microbes	1 870–2 080 (estimated by subtraction)	1 811
Change of organic matter in soil (layer 30 cm deep)	29 880–32 220	1 395 (accumulation)

* After Dąbrowska-Prot et al. (1974), Gołębiowska et al. (1974), Kukielska (1973), Ryszkowski et al. (1973), and unpublished data of Department of Agroecology at Turew, Poland.
† After Satchell 1971.

conditions were used. Since the errors of the methods used are unknown no objective comparison can at present be made.

Conclusions

In conclusion the main features of the energy and matter economy of the types of ecosystem examined can be characterized as follows:

1. The forest ecosystems cover different successional stages up to climax communities and have a very complicated structure. Forests change climatic factors, producing distinct microclimates; they intercept large quantities of incoming energy or water but use them less efficiently than do other ecosystems. Forests have very high capacities for the storage of organic matter in living forms, both in plants and animals. Energy stored in plant biomass can exceed the value of the annual net radiation in the case of rainforest. The ratio of energy accumulated in live and dead organic matter to the cost of rock weathering is highest in forests. The ash contents of plants are lower in forests than in other ecosystems. Thus forests have a less intensive mineral economy, probably because of high retention capacities. In climax forest ecosystems the energy cost of the production of a biomass unit is high since the standing crop is very large.

2. The grassland ecosystems cover different successional stages up to climax communities such as steppe or prairie. They have a simpler structure than forests and possess the highest capacities to store energy in the form of humus. Because of the deep humus layer there is no leaching. They intercept less energy and water than forests but use them more economically. These ecosystems are controlled to a high degree by climatic factors, especially water, in a sequence of dry and wet seasons. Practically all above-ground primary production is shed annually and mineralization processes are intensive. Herbivores act as an important factor in mineral cycling, especially under dry conditions. The use of grasslands for pastures increases the impact of grazing animals which can lead to overexploitation, especially without management of the water regime. In pastures the impact of man on soil formation processes is rather small.

3. Cultivated fields are ecosystems artificially maintained at an early stage of succession with a simple structure, few possibilities of modifying the effect of climatic factors, high productivity, low energy cost in the production of a biomass unit, and with open cycles of mineral circulation. In these ecosystems the costs of maintaining their stability are borne by man. Man influences practically all the ecosystem processes.

References

Assman, E. 1968. Waldertragskunde. PWRL Warszawa (Polish translation).

Briggs, L. G. and H. L. Shantz 1914. Relative water requirement of plants. J. Agric. Res. 3: 1–63.

Burges, A. & F. Raw (ed.) 1967. Soil Biology. Academic Press, London.

Byzova, B. J. 1972. Dykhanie pochvennikh bespozvonochnykh. Ekologya pochvennykh bespozvonochnykh: 3–39. Nauka, Moskva.

Clark, F. E. and E. A. Paul 1970. The microflora of grasslands. Advances in Agronomy 22: 375–435.

Dabrowska-Prot, E., J. Karga and L. Ryszkowski 1974. An attempt to estimate the role of invertebrates in agrocenotic economics. In: Ecological effects of intensive agriculture. L. Ryszkowski (ed.) PWN. Warszawa.

Domracheva, L. I. 1972. Opyt izuchenya biomassy i sezonnoi produkcii pochvennych vodorosley. In: Metody izuchenya i prakticheskogo ispolzovanya pochvennych vodorosley E. A. Shtina (ed.) Kirov.

Duvigneaud, P. (ed.) 1971. Productivity of forest ecosystems. UNESCO. Paris.

Edwards, C. A., D. E. Reichle and D. A. Crossley 1970. The role of soil invertebrates in turnover of organic matter and nutrients. In: Analysis of temperate forest ecosystems. D. E. Reichle (ed.) Heidelberg.

Frenchel, T. 1974. Intrinsic rate of natural increase: the relationships with body size. Oecologia 14: 317–326.

Ghilarov, M. S. 1967. Abundance, biomass and vertical distribution of soil animals in different zones. In: Secondary productivity of terrestrial ecosystems. K. Petrusewicz (ed.) PWN. Warszawa.

Gollerbach, M. M. and E. A. Shtina 1969. Pochvennye vodorosli, Nauka. Leningrad.

Golley, F. B. 1972. Energy flux in ecosystems. In: Ecosystem structure and function. J. A. Wiens (ed.), Oregon Univ. Press, Portland, Oregon.

Golley, F. B., L. Ryszkowski and I. T. Sokur 1975. The role of small animals in temperate forests, grasslands and cultivated fields. In: Small mammals: their productivity and population dynamics. F. B. Golley, K. Petrusewicz & L. Ryszkowski (ed.). Cambridge Univ. Press (in press.)
Golebiowska, J., Z. Margowski and L. Ryszkowski 1974. An attempt to estimate the energy and matter economy in the agrocenosis. In: Ecological effects of intensive agriculture. L. Ryszkowski (ed.) PWN, Warszawa.
Grin, A. M., Y. L. Rauner and V. D. Utekhin 1970. Effektivnost ispolzovanya radiacii i vlagi v lesostepnykh ekosistemakh. Izv. Akad. Nauk SSSR 4: 10–22.
Grodzinski, W. and N. R. French 1974. Production and respiration in populations of small mammals. Trans. 1st Int. Theriol. Congr. Moscow.
Hutchison, V. H., W. G. Whitford and M. Kohl 1968. Relation of body size and surface area to gas exchange in anurans. Physiol. Zool. 41: 65–85.
Kajak, Z. & A. Hillbricht-Ilkowska (ed.) 1972. Productivity problems of fresh waters, Proc. IBP-UNESCO Symp. Kasimierz Dolny, 1970 PWN. Warszawa.
Kleiber, M. 1961. The fire of life. Wiley and Sons, New York.
Kononova, M. 1968. Organicheskoye vieshchestvo pochvy. (Polish translation) PWRL, Warszawa.
Kovda, V. A. 1971. Biosfera i cheloveschestvo. In: V. A. Kovda (ed.) Biosfera i je ressursy. Nauka, Moskva.
Kucera, C. L. and D. R. Kirkham 1971. Soil respiration studies in tall-grass prairies in Missouri. Ecology 52: 912–915.
Kukielska, C. 1973. Primary productivity of crop fields. Bull. Acad. Pol. Sci. Cl. II 21: 109–115.
Kurcheva, G. F. 1971. Rol pochvennykh zhivotnykh v razlozhenni i gumifikacii rastitelnych ostatkov. Nauka, Moskva.
Larin, I. V. 1963. Prirodnyie senokosy i pastvishcha. Selskokhozizdat, Moskva.
Lasiewski, C. R. and W. R. Dawson 1967. A re-examination of the relation between standard metabolic rate and body weight in birds. Condor 69: 13–23.
Lieth, H. 1972. Modeling the primary productivity of the world. Turk Biyoloji Dergisi 22: 85–88.
McNeill, S. and J. H. Lawton 1970. Annual production and respiration in animal populations. Nature 225: 472–474.
Odum, E. P. 1971. Fundamentals of Ecology. Saunders, Philadelphia.
Palmen, E. (ed.) 1971. Proceedings of the IBP meeting on secondary productivity in small mammal populations. Ann. Zool. Fennici 8: 1–185.
Petrusewicz, K. (ed.) 1967. Secondary productivity of terrestrial ecosystems. PWN, Warszawa.
Petrusewicz, K. & A. Macfadyen (eds.), 1970. Productivity of terrestrial animals. Blackwell, Oxford.
Petrusewicz, K. & L. Ryszkowski (eds.) 1969. Energy flow through small mammal populations. PWN, Warszawa.
Phillipson, J. (ed.) 1970. Methods of study in soil ecology. UNESCO, Paris.
Phillipson, J. (ed.) 1971. Methods of study in quantitative soil ecology: population, production and energy flow. Blackwell, Oxford.
Rauner, Y. L. 1972. Teplovi balans rastitelnogo pokrova. Gidrometeoizdat, Leningrad.
Reichle, D. E. (ed.) 1970. Analysis of temperate forest ecosystems. Springer-Verlag, Heidelberg.
Reichle, D. E. 1971. Energy and nutrient metabolism of soil and litter invertebrates. In: Productivity of forest ecosystems. P. Duvigneaud (ed). UNESCO, Paris.

Reichle, D. E., B. E. Dinger, N. T. Edwards, W. F. Harris and P. Sollins 1973. Carbon flow and storage in a forest ecosystem In: Carbon and the biosphere. G. M. Woodwell & E. V. Pecan (ed.) USAEC, Springfield.
Riley, G. A., 1973. Particulate and dissolved organic carbon in the oceans. In: Carbon and the biosphere. G. M. Woodwell & E. V. Pecan (ed.) USAEC, Springfield.
Rodin, L. E. and N. I. Bazilevich 1965. Dynamics of the organic matter and biological turnover of ash elements and nitrogen in the main types of the world vegetation. Nauka, Moscow.
Ryszkowski, L. (ed.) 1974. Ecological effects of intensive agriculture (first attempt at a synthesis) PWN, Warszawa.
Ryszkowski, L. 1974a. Krazenie materii w agrocenozach. Zeszyty Problemowe Postepów Nauk Rolniczych, 155: 19–38.
Ryszkowski, L., J. Goszczyński and J. Truszkowski, 1973. Trophic relationships of the common vole in cultivated fields. Acta Theriologica 18: 125–165.
Samtsevich, S. A. 1965. Active excretions of plant roots and their significance. Fiziol. Rast. 12: 837–846.
Satchell, J. E. 1971. Feasibility study of and energy budget for Meathop Wood. In: Productivity of forest ecosystems. P. Duvigneaud (ed.) UNESCO, Paris.
Shtina, E. A. 1972. Algal biomass in soil and methods of its determination. In: Problems of abundance, biomass and productivity of microorganisms in soil. T. V. Aristovskaya (ed.) Nauka. Leningrad.
Volobuyev, V. R. 1974. Vvedenie v energetika pochvoobrazovanya. Nauka. Moskva.
Wetzel, R. G. and P. H. Rich, 1973. Carbon in fresh water systems. In: Carbon and the biosphere. G. M. Woodwell & E. V. Pecan (ed.) USAEC, Springfield.
Whittaker, R. H. and G. E. Likens, 1973. Carbon in the biota. In: Carbon and the biosphere. G. M. Woodwell & E. V. Pecan (ed.) USAEC, Springfield.
Wiebe, W. J. 1973. Perspectives in microbial ecology. In: Fundamentals of Ecology. E. P. Odum: 484–495.
Winberg, G. G. (ed.), 1968. Metody opredelenya produkcii vodnykh zhivotnykh. Vysheishaya Shkola, Minsk.
Woodwell, G. M. & E. V. Pecan (ed.), 1973. Carbon and the biosphere. USAEC, Springfield.
Zlotin, R. I. and K. S. Hodashova, 1974. Rol zhivotniykh v biologicheskom krugovorte lesostepnykh ekosystem. Nauka. Moskva.

Author's address:

Lech Ryszkowski
Department of Agroecology
64–003 Turew
Kościan
Poland

Factors involved in dynamics of algal blooms in nature

M. Shilo

Introduction

Primary production by algae in the aquatic environment plays a major role in total global production, a role further amplified by the phenomenon of massive algal development reaching dimensions of blooms. These algal blooms and their dynamics therefore are major factors in ecological considerations, becoming increasingly acute due to accelerated eutrophication caused by intensification of agriculture and the ever increasing release of urban and industrial wastes into different water bodies. In order to intervene in, control or manage these blooms we have to know what factors trigger mass algal development, make possible the dominance of certain species and cause their wane and die off. Algal blooms, especially of blue-greens, considerably upset the ecological balance and the quality of waters for human use; especially following the sudden die off, such blooms produce foul taste and odors in waters and cause intoxication of animals and extensive fish mortalities. Perhaps even more important, the mortality and rapid decomposition of algal blooms in nature produce, in turn, depletion in the oxygen content of the water. Proper management of bodies of water must therefore be directed towards the preventions of blooms, on the one hand, and the preservation of established blooms, on the other, to prevent their sudden die off.

For an understanding of the mechanisms of the sudden cyclic mortality of algal blooms, we have centered our own interests on the blue-greens, which are a major component of planktonic algal blooms.

The ecology of blue-green algae

The dominance of blue-green over other algal growth in many aquatic ecosystems seem to be related, first of all, to the unique ability of many blue-green species to tolerate, survive and grow in extreme environmental conditions, and to withstand great fluctuations within short (diurnal) time periods and within limited spaces (such as the surface water layers). The spatial and temporal phenomena are most dramatically exemplified in the continual, heavy blue-green blooms in shallow equatorial lakes (Dunn et al., 1969; Ganf, 1969; Horne & Viner, 1971; Viner, 1969), and in hypersaline ponds or lagoons or in the fish-ponds in Israel during the summer season. These extreme conditions include high temperature (Castenholz, 1969; Allen, 1966), high salt concentrations (Cohen, 1971; Krumbein & Cohen, 1974), high pH, limited light and nutrient concentrations (King, 1970), high H_2S concentrations and low O_2 and even anaerobic conditions (Stewart & Pearson, 1970; Castenholz, 1973).

A second consideration affecting the prevalence of blue-greens is their nutritional versatility. These algae were until recently considered to be obligate photoautotrophs, but it is now clear that many are capable, in addition, of photoheterotrophic and chemoheterotrophic growth. Not only uptake of certain substances, such as acetate, in the light has been demonstrated, but growth depends on various organic substances in dim (Van Baalen et al., 1971) or blue (Pulich & Van Baalen, 1974) light insufficient to allow growth on mineral media, in light on the presence of DCMU which inhibits photosystem II (Rippka, 1972; Stanier, 1973). In CO_2 depleted media (Ingram et al., 1973) and with different sugars chemoautotrophic growth in the dark has been obtained with a number of strains (Rippka, 1972; Miller & Allen, 1972; Pelroy et al., 1972; Hoare et al., 1971).

This nutritional versatility may explain why blue-green algae often become predominant in environments which contain high concentrations of dissolved organic matter; lakes receiving sewage effluents are particularly suitable for their development.

Population fluctuations

From this short survey we can perceive that a considerable body of information has been gathered on the factors involved in mass development and formation of blue-green blooms. Less is known on the equally important and interesting phase of bloom dynamics—the die off of blooms. We have devoted a considerable portion of our research capacities to the elucidation of the agents and conditions involved in the mortality and predation of blue-green algae. Counteracting factors enhancing growth and causing appearance of blooms, we know of continual predations of blue-green by protozoa, inhibition and killing by bacterial antibiotics and lysis by bacteria and cyanophages. The importance of these antagonists cannot yet be fully assessed since we still know little of the extent of their distribution and importance in natural conditions where blue-greens become dominant. In the study of damaging agents, we and several other groups have dealt with the action of cyanophages (Paden & Shilo, 1973; Safferman, 1973; Brown, 1972) and bacteria (particularly Myxobacteria) which lyse algae, some by direct contact only (Shilo, 1970; Daft & Stewart, 1971) and others by producing antibiotic substances (Stewart & Brown, 1971; Mira Shilo, personal communication). Though these agents may not be the principal cause of die off in nature, they might possibly serve as a potential means of control of blooms. The use of cyanophages as control agents for blue-green algal blooms has been suggested by many authors (Goriushyn & Chaplinskyaya, 1966, 1968; Safferman & Morris, 1964; Safferman, 1968; Jackson & Sladecek, 1970); there are major difficulties which have to be overcome for effective economical large scale use.

The high degree of host specificity, the selection of resistant host mutants, and the dependence on environmental factors indicate the complexity of the alga–cyanophage interaction, whose outcome depends not only on inherent properties of both phage and host, but on fluctuations in external conditions as well. Deliberate manipulation, both of viral genetic material and of external conditions, could serve as the key to prospective biological control of undesirable blue-green algal blooms. A possible effective use of this kind of control has been suggested from work in our department with a temperate cyanophage which is induced to become lytic at elevated temperatures (Paden & Shilo, 1973). Such mutants do exist in nature and are being studied at present.

Lethal photooxidation

Here I would like to concentrate on an approach in our work dealing with the lethal photoooxidation to which certain blue-green algae become extremely sensitive under conditions prevailing in blooms.

A characteristic feature observed with highly dense blue-green algal populations, particularly in the upper layers of water where these algae tend to concentrate due to their buoyancy, is the rapid depletion of CO_2 due to photoassimilation and the tremendous oxygen enrichment (up to 200–300% saturation) during active photosynthesis, following a diurnal cyclic pattern. These conditions are especially prevalent in geographical regions with high light intensities and where high oxygen concentrations develop locally in the upper layers of the water. A typical feature of lakes in warm regions is the trapping of the oxygen-rich layer by blocking of gas exchange at the air/water interphase. Our experiments (Abeliovich & Shilo, 1972; Abeliovich et al., 1974) with pure cultures of *Anacystis nidulans* (as well as with *Synechococcus cedrorum*) under conditions of CO_2 depletion showed that these organisms are killed rapidly at physiologically high temperatures (35°) under the photoooxidative conditions. Cells did not die in control tests when illuminated with an oxygen atmosphere containing 5% CO_2, or in a N_2 atmosphere or under O_2 in the dark. The photooxidative effect prevailed and is even more pronounced at a low temperature (4°) even in the presence of CO_2. Finally, the kinetics of the photooxidative effect depend on the growth history of the culture since cells grown first under a CO_2-rich atmosphere resist the photooxidative conditions for a longer period. The role of CO_2 in preventing photooxidative death in *Anacystis* indicates that a protective mechanism (or mechanisms) may be connected with the photosynthetic activity of the cell.

Field experiments

To determine whether photooxidative death can play a role in the sudden die off and decay of blooms in natural water bodies, we tested this in Israeli fishponds where die off of blue-green blooms is commonly observed. Since it was difficult to test this with the natural often mixed blue-green populations in the ponds, we developed a method in which axenic cultures of different blue-green algae were placed in dialysis tubing and suspended in pond waters, and the cells enumerated under different experimental conditions and at different times. These tests completely demonstrated the photooxidative phenomena, described above, in field conditions in both summer and winter. Such experiments, allowing for laboratory type tests under field conditions, demonstrate the likelihood that photooxidation prevails in the natural conditions and that it could well play a key role in the die off phenomenon. Marked differences in sensitivity of different blue-green algal species to photooxidative conditions were found. Resistance was typical for some of the strains, such as *Microcystis*, collected from blooms. Photooxidative death of *Anacystis* cells observed with this system at low winter temperatures in ponds is in accord with the absence of blue-green blooms at low temperatures, and possibly explains the well known seasonal fluctuation of blue-green algae. The sensitivity to photooxidation could well determine the spatial distribution of the blue-green algae and may explain the prevalence of blue-greens in conditions of dim light (as in benthic layers, under stones and sand layers in lakes) and in CO_2 rich waters found in eutrophic conditions.

To understand the mechanism of photooxidative death in blue-green algae, it was interesting to study the levels of superoxide dismutase, an enzyme recently reported to play a key role in the protection of aerobic organisms from the toxic effects of oxygen (Abeliovich et al., 1974). Superoxide dismutase activity began to decrease steadily early in the lag period and by the 6th hour was depleted to 10% of initial activity in *Anacystis* cells shifted from an atmosphere of air to pure oxygen in a medium devoid of CO_2, in which photooxidative death occurs after a lag of 6–8 hours. An additional approach to test the role of superoxide dismutase was to obtain experimental conditions in which cells had different levels of the enzyme and to test their sensitivity to photooxidation. It was found that cells grown under N_2 with 5% CO_2 (i.e. low oxygen conditions) have very low superoxide dismutase levels. When exposed to photooxidative conditions such superoxide dismutaseless cells are most rapidly killed. Such cells rapidly form superoxide dismutase on exposure to air and regain normal resistance to photooxidation, indicating that this enzyme is induced by elevating oxygen concentrations.

Conclusion: possibilities for control

A whole new approach to management of blue-green blooms and their control, based on this knowledge, has now become feasible. First, control may be possible before the bloom reaches peak level by inhibition of photosynthesis (e.g. herbicides) and by the use of agents lowering the superoxide dismutase level to make the algae more sensitive to photooxidation. Second, to maintain established blooms the prevention and protection from photooxidative death could possibly be obtained by raising the superoxide dismutase level of the cells, by enriching with CO_2, by reducing light intensity (especially in wavelengths within the action spectrum of photooxidation) and by the removal through mechanical stirring of the high oxygen-concentration pockets or layers accumulating in the algae-dense surface layer.

Acknowledgments

This study has been supported by a grant from the Deutsche Forschungsgemeinschaft.

References

Abeliovich, A. and Shilo, M. 1972. Photooxidative death in blue-green algae. J. Bacteriol. 111: 682–689.

Abeliovich, A., Kellenberg, D. and Shilo, M. 1974. Effect of photooxidative conditions on levels of superoxide dismutase in *Anacystis nidulans*. Photochem. & Photobiol. 19: 379–382.

Allen, Mary Ann Mennes. 1966. Studies on the properties of some blue-green algae. Ph.D. Thesis, Univ. of California, Berkeley, California.

Brown, Jr., R. M. 1972. Algal viruses. In: Advances in Virus Research. 17: 243–277.

Castenholz, R. W. 1969. Thermophilic blue-green algae and the thermal environment. Bacteriol. Revs. 33: 476–504.

Castenholz, R. W. 1973. Ecology of blue-green algae in hot springs. In: The Biology of Blue-Green Algae. N. G. Carr & B. A. Whitton eds., Blackwell Scientific Publications, Oxford. pp. 379–414.

Cohen, Y. 1971. Studies on the microflora of the Solar Lake. M.Sc. Thesis, The Hebrew University-Jerusalem, Israel.

Daft, M. J. and Stewart, W. D. F. 1971. Bacterial pathogens of freshwater blue-green algae. New Phytol. 70: 819–829.

Dunn, I. G., Burgis, M. J., Ganf, G. G., McGowan, L. M. and Viner, A. B. 1969. Lake George, Uganda: a limnological survey. Int. Ver. Theor. Angew. Limnol. Verh. 17: 284–288.

Ganf, G. G. 1969. Physiological and ecological aspects of the phytoplankton of Lake George, Uganda. Ph.D. Thesis, University of Lancaster, U.K.

Goriushyn, V. A. and Chaplinskyaya, S. M. 1966. Existence of viruses of blue-green algae. Mykrobiol. Zhurn. (Kiev) 28: 94–97.

Goriushyn, V. A. and Chaplinskyaya, S. M. 1968. The discovery of viruses lysing blue-green algae in the Dneprovsk reservoirs. Tsvetenie Vody. Dumka, Kiev.

Hoare, D. S., Ingram, L. O., Thurston, E. L. and Walkup, R. 1971. Dark heterotrophic growth of an endophytic blue-green alga. Arch. Mikrobiol. 78: 310–321.

Horne, A. J. and Viner, A. B. 1971. Nitrogen fixation and its significance in tropical Lake George, Uganda. Nature Lond. 232: 417–418.

Ingram, L. O., Van Baalen, C. and Calder, J. A. 1973. Role of reduced exogenous organic compounds in the physiology of the blue-green bacteria (algae): Photoheterotrophic growth of an 'Autotrophic' blue-green bacterium. J. Bacteriol. 114: 701–706.

Jackson, D. F. and Sladecek, V. 1970. Algal viruses–eutrophication control potential. Yale Sci. Mag. 44: 16–22.

King, D. L. 1970. The role of carbon in eutrophication. J. Water Poll. Control Fed. 42: 2035–2051.

Krumbein, W. E. and Cohen, Y. 1974. Biologen, klastische und evaporitische sedimentation in einem mesothermen monomiktischen unfernahen See (Golf von Aqaba). Geol. Rdsch. 63: 1035–1065.
Miller, J. S. and Allen, M. M. 1972. Carbon utilization patterns in the heterotrophic blue-green alga *Chlorogloea fritschii*. Arch. Mikrobiol. 86: 1–12.
Padan, E. and Shilo, M. 1973. Cyanophages—Virus attacking blue-green algae. Bacteriol. Revs. 37: 343–370.
Pelroy, R. A., Rippka, R. and Stanier, R. Y. 1972. Metabolism of glucose by unicellular blue-green algae. Arch. Mikrobiol. 87: 303–322.
Pulich, Jr., W. M. and Van Baalen, C. 1974. Growth requirements of blue-green algae under blue light conditions. Arch Microbiol. 97: 303–312.
Rippka, R. 1972. Photoheterotrophy and chemoheterotrophy among unicellular blue-green algae. Arch. Mikrobiol. 87: 93–98.
Safferman, R. S. and Morris, M. E. 1964. Control of algae with viruses. J. Amer. Water Works Ass. 56: 1217–1224.
Safferman, R. S. 1968. Virus diseases in blue-green algae. In: Algae, man and the environment. Daniel F. Jackson ed., Syracuse University Press, New York.
Safferman, R. S. 1973. Phycoviruses. In: The Biology of blue-green algae. N. G. Carr & B. A. Whitton eds., Blackwell Scientific Publications, Oxford, pp. 214–237.
Shilo, Miriam. 1970. Lysis of blue-green algae by myxobacter. J. Bacteriol. 104: 453–461.
Stanier, R. Y. 1973. Autotrophy and heterotrophy in unicellular blue-green algae. In: The Biology of blue-green algae. N. G. Carr & B. A. Whitton eds., Blackwell Scientific Publications, Oxford, pp. 501–518.
Stewart, J. R. and Brown, R. M. 1971. Algicidal non-fruiting myxobacteria with high G + C ratios. Arch. Mikrobiol. 80: 176–190.
Stewart, W. D. P. and Pearson, H. W. 1970. Effects of aerobic and anaerobic conditions on growth and metabolism of blue-green algae. Proc. R. Soc. B. 175: 293–311.
Van Baalen, C., Hoare, D. S. and Brandt, E. 1971. Heterotrophic growth of blue-green algae in dim light. J. Bacteriol. 105: 685–689.
Viner, A. B. 1969. The chemistry of the water of Lake George, Uganda. Int. Ver. Theor Angew. Limnol. Verh. 17: 289–296.

Author's address:

M. Shilo
Department of Microbiological Chemistry
The Hebrew University-
Hadassah Medical School
Jerusalem
Israel

Discussion

Summarized by Th. Alberda

Participants: the authors O. W. Heal (U.K.), H. Lieth (U.S.A.), S. F. MacLean Jnr (U.S.A.), M. Shilo (Israel), L. Ryszkowski (Poland), F. E. Wielgolaski (Norway), together with M. Alexander (U.S.A.), W. Grodzinski (Poland), A. B. Hillbricht-Ilkowska (Poland), E. Inoue (Japan), W. E. Krumbein (F. R. Germany), J. Sidorowicz (Zambia), C. O. Tamm (Sweden), H. Veldkamp (The Netherlands), D. F. Westlake (U. K.), R. G. Wiegert (U.S.A.).

The discussion on these four papers concentrated mainly on the global primary production patterns, as calculated by Lieth, and the magnitude of some efficiency values in calculating secondary productivity, as presented by Heal and MacLean.

As to the productivity maps presented by Lieth, *Tamm* remarked that those maps, although excellent by themselves, may appear to be too good in the sense that the values presented for different parts of the globe may have the tendency to become final. For the tundra area at any rate there is considerable variation in productivity and he considers these variations more interesting and of more importance than the values calculated by Lieth's regression functions, as they give an idea of the range in which man might improve or deteriorate natural production rates.

Sidorowicz thought that the productivity values for Central Africa and Madagascar were slightly overestimated as calculated from the length of the growing period. He thought factors like quality of soils and amount of rainfall more important than the length of the growing season.

Another aspect of primary production on earth was mentioned by *Krumbein*, who pointed out that back into geological history there have been completely different overall productivity patterns, leading to the accumulation of large amounts of fossil energy. A closer study of the succession of ecosystems could probably add useful information to the present day productivity calculations.

Finally *Inoue* asked whether the model presented would still be quantitatively valid at the end of this century.

Lieth answered that the model serves to calculate primary productivity on our globe from climatic parameters and to compare these with actual values. Of course the calculated values must not be considered as final, and he agreed with both Tamm and Sidorowicz that variations and their possible cause as well as factors like soil fertility and rainfall patterns should be taken into consideration. However, as there is an urgent need for overall data on today's productivity, the model was devised and presented as soon as possible. In general, the model as such can serve the purpose, even up to the end of this century. However, it can be refined and extended in the future, as has been done in the past, and the latest model (the Hague model) gives values that already show lower productivities for Central Africa than did the preceding ones.

The Hague model also allows comparative predictions about the productivity level of managed and 'natural' ecosystems across the world. If the trend established in North America holds true for the entire world, one can conclude that agricultural systems in warmer humid regions are less productive than the natural vegetation they replace. If this remains so, in the future we can expect the primary productivity of the continents to drop by 25–40%. The lower productivity is caused by crop selection for cash value rather than optimal utilization of the vegetative period. In North Carolina we found, for example, productivity values for cotton of 1–2 t/ha, tobacco 3–7 t/ha, wheat 4–14 t/ha, against the calculated natural potential for forests of 7–19 t/ha. Even subtropical crops like sweet potatoes 5–13 t/ha, and corn 6–14 t/ha reached on average only one half to two thirds of the potential value.

Concerning the growth efficiency of saprovore microrganisms discussed by Heal and MacLean, several people (*Alexander*, *Veldkamp* and *Hillbricht-Ilkowska*) felt that some of the values recorded in the literature were too high, particularly the generalized value of 0.60 derived by Payne (1970). These were mainly derived from batch cultures with bacteria growing at maximum rates. In studies using continuous cultures and in nature such high growth rates are probably not achieved. *Heal* and *MacLean* pointed out that in their calculations a value of 0.4 was used instead of 0.6. However, in the literature a broad range of values may be found. Recent fungal studies, which included estimates of losses, give values of 0.25–0.50. Field estimates, using ^{14}C substrates in soil (McGill et al, 1973; Shields et al, 1973), in fresh water (Hobbie & Crawford, 1969) and estuaries (Williams, 1973) also shows high yield coefficients. These data were the basis for the use of the factor 0.4. One asset of the model is that it allows us to explore the consequences of alternative systems.

Hillbricht-Ilkowska mentioned her experience with aquatic ecosystems in which the intake efficiency I_n/NP_{n-1} of filter feeders feeding on green algae of sizes lying within their size selection capacity, can reach efficiencies close to 1.0. Also, when not consumed by filter feeders the amount of consumed mass derived directly or indirectly from algae can form about 30% of their production.

Heal and *MacLean* answered that it would be very interesting to apply their method to marine and fresh-water ecosystems. They are not sufficiently familiar with the literature on aquatic ecosystems to attempt to generalize, but it seems that there exist consistent differences affecting heterotrophic productivity.

Grodzinski commented on the assimilation efficiency of vertebrate herbivores, presented as 0.50 in Heal and MacLean's Table 1, page 91. He pointed out that small homeotherms have efficiencies in the range of 0.7–0.9, which he supposed to be necessary because of their high energy demand, and he suggested the separation of the vertebrate herbivores into poikilotherms and small and large homeotherms. The authors answered that the herbivores can certainly be subdivided in the model, but that graminivorous-passerines, murid rodents etc. which possess the high assimilation efficiencies referred to, were not included in the model on account of the belief that they only harvested a small fraction of the net primary production. Another feature mentioned by Grodzinski, viz. that ruminants usually have lower assimilation efficiencies than small monogastrics, is contradicted by data from White et al. and Batzli that indicate that reindeer and caribou assimilate a good deal more from tundra vegetation than do brown lemmings.

Westlake referred to the fact that, according to the kind of ecosystem examined, primary production is either accumulated or completely respired. He asks himself what meaning must be given to secondary productivity unless the system is stopped at some step, such as a yield of an economically useful organism or, more generally, of the food of one species. How otherwise is it possible to make any generalizations from the number of reorganizations involved or the mean retention time of an organic molecule?

Heal and *MacLean* answered that their concept of productivity is certainly not equivalent to yield; it represents more nearly the number of biological reorganizations between primary production and, eventually, respiration and release of energy in radiation. One interesting use of the model is in the prediction of this pattern of successive reorganizations. Clearly, however, yield to man precludes recycling within the saprovore-based system, and thus total apparent productivity is reduced by an amount that can be greater than the amount harvested.

Finally *Wiegert* mentioned that the total secondary production by microbial saprovores, as calculated by Heal and MacLean, could be greater than the input. This led him to discuss the definition of secondary production. Although he is aware of the difficulty of measuring the successive turnover levels in a saprovore system, he suggested defining the secondary production as the amount of protoplasm leaving a particular secondary level and serving as an input to organisms of the following level.

Heal and *McLean* agreed that the term secondary productivity as the sum of all heterotrophic productivity is certainly misleading. The distinction between $2^\circ, 3^\circ, 4^\circ, \ldots, n^\circ$ production-consumption level is a useful one conceptually, and as such it is incorporated in the model; however, where recycling within the saprovore trophic system exists, these concepts are difficult or impossible operationally. Recycling within the saprovore system is reflected in the observed fluctuations in microbial populations in soil. Estimates of microbial production and generation time based on field population changes are comparable with the concept of productivity used in the paper.

Session 3

Diversity, stability and maturity in natural ecosystems

Chairman: V. Westhoff

Diversity, stability and maturity in natural ecosystems

Gordon H. Orians

Introduction

The belief that natural ecosystems become more diverse and, hence, more stable with time after a disturbance is widely accepted and regularly repeated in ecology textbooks (Clements & Shelford, 1939; Colinvaux, 1973; Collier et al., 1973; Odum, 1953). There are suggestions on empirical and theoretical grounds of quantitative relationships between diversity and some measure of stability (Hairston et al., 1968; Hurd et al., 1971; Goel et al., 1971; Leigh, 1965; May, 1972, 1973b; Murdoch et al., 1972; Paine, 1969; Patten, 1963; Pimentel, 1961; Volterra, 1937; Watt, 1964) but the correlations, not to mention causations, are still obscure. In any case, the popularity of the notions that succession generates diversity and that diversity enhances stability predates empirical or theoretical justification. Also, the concepts are normally discussed with poorly defined terms, reflecting an uncertainty about what concept(s) of stability are useful in ecology and, even more important, what we wish to understand about natural ecosystems.

Unfortunately it will be difficult to establish causal relationships between stability and diversity because we must measure one or more concepts of stability in ecological systems differing only in some measure of diversity. The easiest ecosystems to compare cannot provide adequate proof because differences in diversity are usually associated with differences in the physical environment and other complicating factors. Rather than demonstrating greater stability in species-rich systems we may only be showing that species are more vulnerable to disturbances in marginal environments, or that environmental constancy facilitates diversity while reducing perturbations that might affect stability. In fact, we are confronted with the apparent paradox that stability in natural ecosystems seems to be associated with diversity whereas increasing the diversity of a variety of model ecosystems tends to reduce rather than enhance their stability (May, 1973b).

Another attractive option, the perturbation experiment, also confronts serious interpretational problems because the species in ecosystems are coevolved and removal or addition of one or more of the species not only changes the diversity of the system but many of the interaction parameters as well. Which of the changes should be attributed to differences in diversities and which should be attributed to changed interaction patterns is difficult to determine. If the species are allowed to adjust evolutionarily to the new association patterns, the final stabilities may be very different from the ones observed immediately after the perturbation.

Theoretical developments face similar problems. Most formal analyses of concepts of stability are based on non-linear population equations. Given a system of these equations, equilibrium populations are determined by setting all growth rates equal to zero and then analyzing the effects of perturbations around the equilibrium. A common tool is the use of an $m \times m$ matrix, referred to as the *interaction matrix*, each element of which describes the effect of species j on species i near equilibrium. Analysis of an interaction matrix reveals whether or not the system is stable, i.e. if it returns to its original state after a perturbation, and the speed of return, which can be estimated from the values of the elements. Most theoretical studies have focussed on local stability under deterministic environments and small perturbations. It is, however, also possible to construct matrices whose elements are random variables and to analyze responses to stronger perturbations (May, 1973a), or to consider qualitative matrices in which only the signs of the interactions are known and not their values.

In all analyses to date, however, the properties of the species in the system have been taken as given and not permitted to change with time. Therefore, behavioral and physiological changes in response to the direct and indirect effects of the perturbations, as well as evolutionary changes in species characteristics, are not included. Also, adaptive considerations have not entered into the decisions about what properties the species are given initially. These limitations are not trivial since rapid evolution does occur in response to perturbations and the properties of species are not randomly determined.

Yet another shortcoming of current theory is that it deals primarily with variations in time and not in space and the analytical problems in dealing with *both* time *and* space are formidable. There are some pioneering studies of patterns in space in natural (Aubreville, 1938; Watt, 1947) and artificial (Huffaker, 1958) communities, and recently, theoreticians have turned their attention to spatial heterogeneity. Results to date suggest a very significant role for spatial heterogeneity for several stability properties of ecological systems (Andrewartha & Birch, 1954; Cohen, 1970; Horn & MacArthur, 1972; Janzen, 1970; Levin, 1974; Levins & Culver, 1971; Roff, 1974a, b) but such analyses have only begun to deal with the real complexities of spatial variations in environmental quality.

In this paper I review some of the ways in which interactions at the level of populations can influence the number of species living together and consider how these interactions affect stability properties of ecological systems that may be of theoretical and practical interest. Recognizing that selection acts to maximize fitness of individuals and not directly on the stability properties of the system, I concentrate on how the physical environment and biological interactions mold individual species adaptations. This analysis sheds some light, however feeble, on the problem of which kinds of questions are most usefully posed about stability, how perturbations can be measured meaningfully and how theoretical evolutionary ecological studies can provide insights into community level properties.

I use the term *richness* to refer to the total number of species living together, and the term *diversity* when the species are weighted according to their abundance, size, energy flow or some other measure of presumed importance. I do not use the term maturity because I am unable to define it in any useful or rigorous way. Since the term 'stability' has been and is applied to a number of different concepts, I begin by distinguishing and naming these concepts in an effort to avoid, or at least reduce, arguments that are semantic rather than substantive.

The meanings of stability

The concept of stability usually refers to the tendency of a system to remain near an equilibrium point or to return to it after a disturbance. These meanings can conveniently be described using familiar terms from physics and illustrated graphically (Fig. 1) with stable points in a phase space (Lewontin, 1969).

1. *Constancy*—a lack of change in some parameter of a system, such as the number of species, taxonomic composition, life form structure of a community, or feature of the physical environment. The term carries no causal connotations, as for example its use in a statement that climates are more stable (constant) in the tropics. Similarly, the expression that lemming populations are unstable (inconstant) refers to the magnitude of fluctuations in density, implying nothing about causation.
2. *Persistence*—the survival time of a system or some component of it. In this sense one population might be considered more 'stable' than another if its mean time to extinction were longer (Roff, 1974a,b). Leigh (1965) analysed the relationships between this concept of stability and productivity and diversity.
3. *Inertia*—the ability of a system to resist external perturbations. Changes can be measured in a variety of ecologically interesting parameters, including but not limited to those listed above under constancy. This is the meaning of stability in MacArthur's (1955) analysis of food web structure where he pointed out that with certain food webs, equal changes in the abundance of one of the species would produce different changes in the abundances of the others. It is similar to Holling's (1973) concept of resilience.
4. *Elasticity*—the speed with which the system returns to its former state following a perturbation. A measure of elasticity is contained in a community matrix and, hence, at the formal level the possibility of comparing elasticities in a quantitative way exists.
5. *Amplititude*—the area over which a system is stable. A system has a high amplitude if it can be considerably displaced from its previous state and still

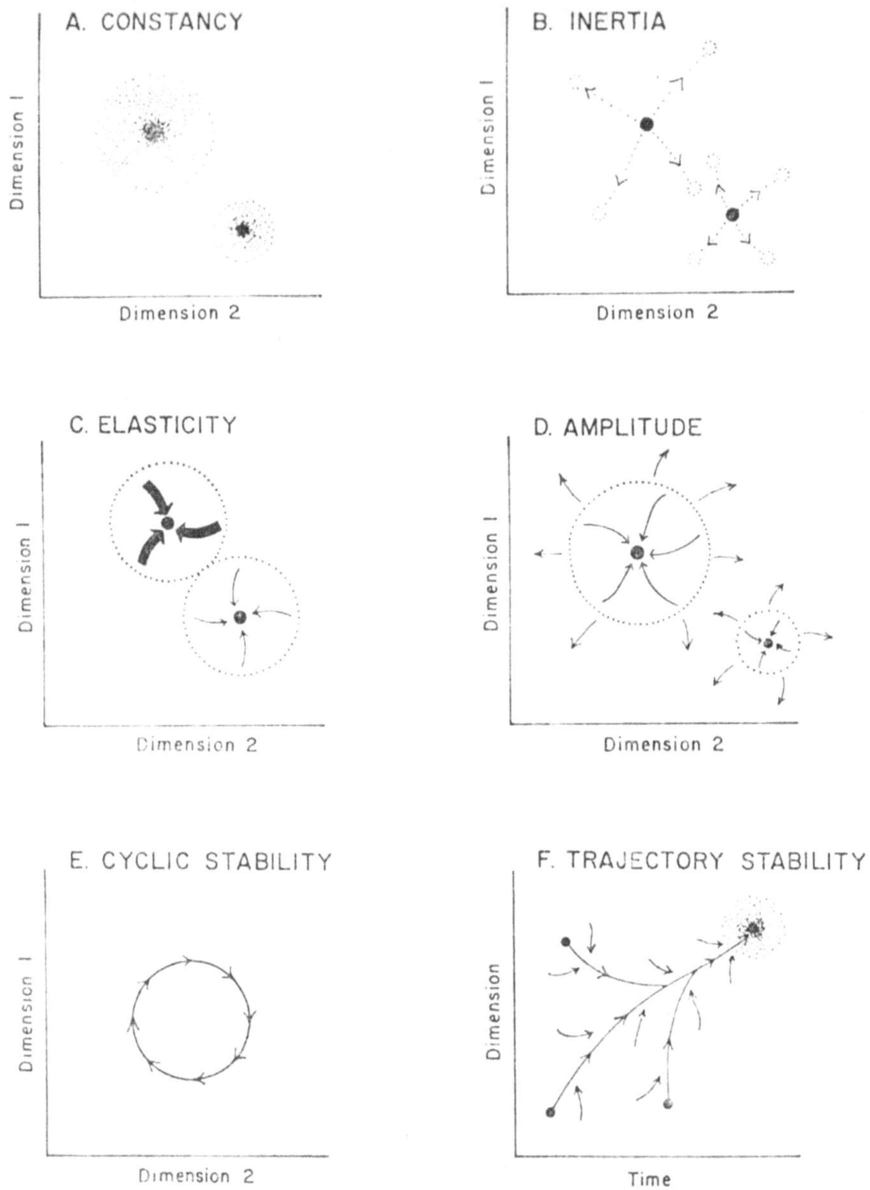

Figure 1. Graphic representations of some concepts of stability.

A. Constancy. The frequency of occurrence of states of the systems is indicated by the intensity of stippling.

B. Inertia. The dotted circles represent states of the systems following perturbations indicated by the arrows.

C. Elasticity. The speed of return is proportional to the thickness of the arrows.

D. Amplitude. The dotted lines enclose the regions within which the systems return to their initial states following disturbances.

E. Cyclic Stability. The stable limit cycle might also be represented by a doughnut indicating the probability of finding the system at a particular state.

F. Trajectory Stability. The system converges to a particular state from a variety of starting positions.

return to it. This meaning of stability is sometimes referred to as *global stability*, and is of particular interest for a number of applied ecological problems.

6. *Cyclical Stability*—the property of a system to cycle or oscillate around some central point or zone. Some important ecological interaction processes, notably predator-prey systems, have this property, a pattern referred to as a *stable limit cycle* (May, 1972).

7. *Trajectory Stability*—the property of a system to move towards some final end point or zone despite differences in starting points. This is the meaning of stability during plant succession where a single 'climax' state may be reached from a variety of starting points.

This listing of the meanings attached to the concept of stability is not intended as a classification system because the terms are not comparable. Constancy and persistence are descriptive terms implying nothing about underlying dynamics. Cyclic and trajectory stability have measures of inertia, elasticity and amplitude associated with them, etc. The separation of concepts is presented only to illustrate the many meanings of stability, the existence of which presumably reflects a need for a variety of notions relating to fluctuations. None of them, however, has any intrinsic meaning except with reference to some particular question we wish to ask about ecological systems. Except for simple 'how' questions about constancy, most of the questions ecologists are interested in relate some measure of stability to a perturbation. For these relationships to be insightful, perturbations should be related to the evolutionary histories of the organisms experiencing the perturbations, and measured in terms of the total investments that must be made to increase or maintain fitness during those perturbations. Therefore, I now turn to a consideration of the evolutionary responses of organisms to perturbations caused by the physical environment, competitors and predators.

Evolutionary responses to changing physical environments

Since the physical environment is the basic independent variable influencing the evolution of organisms, its relative constancy is an important determinant of the characteristics of organisms and how they respond to perturbations. It is usually easy to determine optimal phenotypes in constant environments but difficult to do so in fluctuating environments (Levins, 1968). A general theory of responses to fluctuating environments must include the following four factors: (1) the proportion of time that a given environmental state exists relative to the life cycle of an organism, that is how often and for what lengths of time is the organism confronted with that environment; (2) the effects of an environmental state on fitness; (3) the total investment required to improve fitness by one unit in that environment (includes energetic costs and losses in fitness in other states); and (4) the predictability and regularity of occurrence of a state. We assume that adaptation to one environmental state (or ecological task) requires lowered ability in handling other states or tasks. If it were possible to be a 'master of many trades' there would presumably be far fewer trades (species) in the world.

Responses to variability in resource availability

The theory of resource harvesting has received considerable attention in recent years and several versions of optimal foraging theory exist (Charnov, 1973; MacArthur, 1972; MacArthur and Pianka, 1966; Paloheimo, 1971; Pearson, 1974; Schoener, 1971). All are concerned with the decisions a predator should make to maximize energy intake per unit of time spent foraging, and the conclusions are similar, at least for predators that encounter and handle prey individually and do not pursue dangerous prey.

Stated without formal proof here, these theories suggest the following rules for expectations of resource harvesting characteristics of organisms: (1) prey should be added to or dropped from the diet independently of their own abundances but dependent only on the abundances of higher ranked prey types (those that yield more energy per unit time); (2) phenotypes specialized for harvesting particular resources are most likely to occur for resources that are common, reliable and sufficiently unique that their efficient use requires specific phenotypic adaptations; (3) specialization to microforaging sites should be restricted to sites providing reliable resources; (4) given equal resource constancy, specialization should be positively correlated with productivity of the environment (encounter rates with prey) if there is competition for resources.

There are also theoretical reasons for expecting a limit to the similarity of two organisms that compete for resources. If the resource base is absolutely constant there is no theoretical limit to the similarity of competing species but even small fluctuations reduce the amount of overlap to a much lower value that is relatively insensitive to the absolute amount of environmental variability (May & MacArthur, 1972). This conclusion is not dependent on the specific form of competition equation used or the exact form of the resource harvesting curves of the competing species and is, therefore, presumably robust. Since most natural environments have substantial resource variability, an important corollary is that differences in community richness are caused primarily by differences in the total range of resources and the degree of specialization of the component species, but not by differences in the amount of overlap in resource utilization.

Responses to changes in predation pressures

The lag in responses of predators to changes in the abundances of their prey causes oscillations which are more likely to be stable if the predator has a high prey threshold (Rosenzweig & MacArthur, 1963; Rosenzweig, 1973), if the predator is prevented from becoming more common for reasons other than prey abundance (territoriality, pressure from its own predators, etc.), or if the prey are patchily distributed. The importance of environmental heterogeneity has been demonstrated in the laboratory (Huffaker, 1958) and the field (Tahvanainen & Root, 1972) and is expressed in some theoretical treatments of predation (Andrewartha & Birch, 1954; Janzen, 1971). Since prey are almost always patchily distributed in nature, spatial heterogeneity is probably the most important generator of cyclical stability in predator-prey systems in nature.

Predators also influence the properties of ecosystems through their evolutionary effects on the characteristics of their prey. An idea with profound implications for community structure concerns the role of predators in affecting the diversity in appearance of their prey (Rand, 1967). A predator foraging in a featured environment with an array of prey species finds individuals of each prey species with different ease (Smith, 1972). In particular, those individuals differing most in such features as size, shape, color, background mimicry, hiding place, escape responses, etc. are less likely to be captured than those more similar to individuals of other species. Such selection, manifest over many generations, should produce a divergence of the prey. In conjunction with the amount of physical complexity of the environment, this may determine how many species of prey can coexist (Ricklefs & O'Rourke, in press).

Community consequences of individual strategies

From this very brief review of the theory of adaptation to varying environments, the machinery of resource harvesting and competition and predator prey interactions, we can conjecture how they will affect persistence, constancy, inertia, elasticity, amplitude, cyclical stability and trajectory stability (Table 1). Factors decreasing these stabilities are generally the inverse of those increasing them and are not listed. Several conclusions that might tentatively be drawn from this table are:

(1) Some types of stability, e.g. inertia and cyclical stability, or elasticity and trajectory stability are generally affected similarly;

(2) inertia depends strongly on temporal and spatial heterogeneity;

(3) elasticity and amplitude are differently affected by several factors and may often be inversely correlated;

(4) species that evolved in physically constant environments and/or in ecosystems of high inertia will perceive as serious perturbations those that are perceived as minor ones by species with different evolutionary histories.

These conclusions are relevant to a comparison of the properties of tropical (the most diverse and often regarded as the most stable) and temperate forests (relatively species poor). A concise summary of features of the organisms living in these two ecosystems is presented in Table 2. Tropical forests probably have relatively high inertia because of high species richness (prey heterogeneity and multiplicity of energy pathways) but may have low elasticity and amplitude

Table 1. Environmental factors and phenotypic characteristics of species that increase different kinds of stability.

A. Persistence
   1. Environmental heterogeneity in space and time
   2. Large patch sizes
   3. Constant physical environment
   4. High resource utilization thresholds of predators

B. Inertia
   1. Environmental heterogeneity in space and time
   2. Greater phenotypic diversity of prey
   3. Multiplicity of energy pathways
   4. Intraspecific variability of prey
   5. High mean longevity of individuals of component species (Frank, 1968)

C. Elasticity
   1. High density-dependence in birth rates
   2. Short life cycles of component species
   3. Capacity for high dispersal
   4. Strong migratory tendencies
   5. Generalized foraging patterns

D. Amplitude
   1. Weak density-dependence in birth rates
   2. Intraspecific variability of component species
   3. Capacity for long-distance dispersal
   4. Broad physical tolerances
   5. Generalized harvesting capabilities
   6. Defense against predators not dependent on a narrow range of hiding places

E. Cyclic Stability
   1. High resource-utilization thresholds
   2. Long lag times in response of species to changes in resource availability
   3. Heterogeneity of environment in space and time

F. Trajectory Stability
   1. Strong organism-induced modifications of the physical environment
   2. All factors increasing elasticity.

Table 2. General adaptive characteristics of species in temperate and tropical forests.

Temperate forests	Tropical forests
1. Low species richness in most taxa	1. High species richness in most taxa
2. Proportionally fewer rare species	2. Proportionally more rare species
3. High reproductive rates	3. Low reproductive rates
4. Extensive seed dormancy	4. Little or no seed dormancy (Gómez-Pompa et al., 1972)
5. High average dispersal rates	5. Low average dispersal rates
6. Many migratory species, short barriers ineffective	6. Few migratory species, short barriers effective (Diamond, 1973a, b, MacArthur, Diamond & Karr, 1972)
7. More habitat generalists	7. Fewer habitat generalists
8. More dietary generalists?	8. More dietary specialists?

because of low density-dependence of birth rates, habitat specializations of many species, no seed dormancy, low average dispersal rates, and few migrants. Moreover, some perturbations, such as creating single species stands of plants (the standard agroecosystem), are clearly more hazardous in tropical regions (Gómez-Pompa et al., 1972; Janzen, 1973). Therefore we can expect high extinction rates among tropical species as a result of converting tropical forests to small 'islands' whereas similar treatment of temperate forests caused relatively few extinctions. Many types of perturbations that have had minor effects in temperate zones may be more significant in the tropics where they represent a more radical departure from the evolutionarily typical ones.

Conclusions

The previous analysis suggests that any attempts to find general relationships between diversity and 'stability' are likely to be fruitless. The questions posed in this important area of inquiry must be much more sharply focused so that the type of stability of interest is clearly specified and its relevance to underlying causal processes directly postulated. 'May's Paradox,' that in nature stability (constancy) is apparently associated with diversity (species richness) suggests that ecosystem behavior in response to perturbations depends primarily on the adaptive characteristics of the organisms in the system. Since these characteristics reflect past histories of experiences with perturbations and the continual evolution of associated species, we need to understand these adaptations better if we wish to improve our predictive powers about the effects of perturbations.

Fitness benefits are characteristic of virtually all the adaptations of organisms. Some traits, such as behavioral ones, may be more flexible than others, such as morphological ones, but compromises are unavoidable. If we know the total investment per unit of fitness improvement for a set of perturbations and if we know the extent to which species in an ecological system are adapted to these perturbations, we should be in a better position to predict the effects of new and unusual perturbations. Studies of the traits of organisms that have been exposed to different kinds and frequencies of perturbations will be especially helpful in making these predictions. If field ecologists can gather this kind of information, it may then be used in more formal analyses of interaction matrices, thereby overcoming what are probably the most serious weaknesses in current mathematical models of ecosystem behavior. These models would thus fall between the purely general ones that now characterize community ecology and the very specific ones commonly used in management that attempt to specify in detail the properties of one species of interest. This intermediate level of analysis may prove to be very powerful in solving the ecosystem problems posed by the severe and extensive perturbations resulting from the unprecedented activities of human beings.

Currently, ecological advice on matters relating to community stabilities is highly intuitive and is not very different from Aldo Leopold's dictum that the first rule of intelligent tampering is to save all the pieces. Since we do not yet know the consequences of the loss of pieces (species) the best strategy may in fact be to make every effort to save all the pieces and the environments they need. Nevertheless, scientific ecology should be able to provide better advice to decision makers who are responsible for difficult choices in matters relating to preservation, planning perturbations, etc. For this vital need the wedding of evolutionary ecology and formal community analysis holds promise of powerful insights.

Acknowledgments

Useful comments were provided by several members of the faculty of the Department of Biology, Washington University, St. Louis where the ideas developed in this paper were first presented. Subsequent improvements were greatly aided by suggestions from Simon Levin, Gilberto Gallopin and W. T. Edmondson. Thinking time was provided by the John Simon Guggenheim Memorial Foundation and sabbatical leave salary from the University of Washington, Seattle.

References

Andrewartha, H. G. and L. C. Birch. 1954. The Distribution and Abundance of Animals. Univ. of Chicago Press.

Aubreville, A. 1938. La Forêt coloniale; les forêts d'Afrique équatoriale. Bois & For. Trop. 2: 24–35.

Charnov, E. L. 1973. Optimal Foraging: some Theoretical Considerations. Unpublished Ph. D. Thesis, Univ. of Washington.

Clements, F. E. and V. E. Shelford. 1939. Bio-ecology. John Wiley & Sons, New York.

Cohen, J. E. 1970. A Markov contingency table model for replicated Lotka-Volterra systems near equilibrium. Amer. Nat. 104: 547–559.

Colinvaux, P. A. 1973. Introduction to Ecology. John Wiley & Sons, New York. 621 pp.

Collier, B. D., G. W. Cox, A. W. Johnson and P. C. Miller. 1973. Dynamic Ecology. Prentice-Hall, Englewood Cliffs, N. J. 563 pp.

Diamond, J. M. 1972. Comparison of faunal equilibrium turnover rates on a tropical and a temperate island. Proc. Nat. Acad. Sci. 68: 2742–2745.

Diamond, J. M. 1973. Distributional ecology of New Guinea birds. Science, 179: 759–769.

Frank, P. W. 1968. Life histories and community stability. Ecology, 49: 355–357.

Goel, N. S., S. C. Maitra and E. W. Montroll. 1971. On the Volterra and other nonlinear models of interacting populations. Rev. Mod. Phys. 43: 231–276.

Gómez-Pompa, A., C. Vásquez-Yanes, and S. Guevara. 1972. The tropical rain forest: a non-renewable resource. Science, 177: 765–769.

Hairston, N. G., J. D. Allen, R. K. Colwell, D. J. Futuyma, J. Howell, M. D. Lubin, J. Mathias and J. H. Vandermeer. 1968. The relationship between species diversity and stability: an experimental approach with protozoa and bacteria. Ecology, 49: 1091–1101.

Holling, C. S. 1973. Resilience and stability of ecological systems. Ann. Rev. Ecol. & Systematics, 4: 1–23.

Horn, H. S. and R. H. MacArthur. 1972. On competition in a diverse and patchy environment. Ecology 53: 749–752.
Huffaker, C. B. 1958. Experimental studies on predation: dispersion factors and predator-prey oscillations. Hilgardia, 27: 343–383.
Hurd, L. E., M. V. Mellinger, L. L. Wolf, and S. J. McNaughton. 1971. Stability and diversity at three trophic levels in terrestrial ecosystems. Science 173: 1134–1136.
Janzen, D. H. 1970. Herbivores and the number of tree species in tropical forests. Amer. Nat. 104: 501–528.
Janzen, D. H. 1971. Seed predation by animals. Ann. Rev. Ecol. Syst., 2: 465–492.
Janzen, D. H. 1973. Tropical agroecosystems. Science, 182: 1212–1219.
Leigh, E. G. 1965. On the relationship between productivity, biomass, diversity and stability of a community. Proc. Nat. Acad. Sci., 53: 777–783.
Levin, S. A. 1974. Dispersion and population interactions. Amer. Nat. 108: 207–228.
Levins, R. 1968. Evolution in Changing Environments. Princeton Univ. Press, Princeton, N.J.
Levins, R. 1969. Some demographic and genetic consequences of environmental heterogeneity for biological control. Bull. Entomol. Soc. Amer. 15: 237–240.
Levins, R. and D. Culver. 1971. Regional coexistence of species and competition between rare species. Proc. Nat. Acad. Sci. 68: 1246–1248.
Lewontin, R. C. 1969. The meaning of stability. In: Diversity and stability in ecological systems. Brookhaven Natn. Lab. Springfield, Va. Symp. Biol. No 22: 13–24.
MacArthur, R. H. 1955. Fluctuations of animal populations, and a measure of community stability. Ecology, 36: 533–536.
MacArthur, R. H. 1972. Strong or weak interactions? Trans. Conn. Acad. Arts & Sciences, 44: 177–188.
MacArthur, R. H. 1972. Geographical Ecology. Harper & Row.
MacArthur, R. H. and E. R. Pianka. 1966. On optimal use of a patchy environment. Amer. Nat. 100: 603–609.
MacArthur, R. H., J. M. Diamond and J. R. Karr. 1972. Density compensation in island faunas. Ecology, 53: 330–342.
May, R. M. 1972. Limit cycles in predator-prey communities. Science, 177: 900–902.
May, R. M. 1973a. Stability in randomly fluctuating versus deterministic environments. Amer. Nat. 107: 621–650.
May, R. M. 1973b. Stability and complexity in model ecosystems. Princeton Monographs in Population Biology No. 6. Princeton Univ. Press, Princeton, N.J. 235 pp.
May, R. M. and R. H. MacArthur. 1972. Niche overlap as a function of environmental variability. Proc. Nat. Acad. Sci. 69: 1109–1113.
Murdoch, W. W., F. C. Evans, and C. H. Peterson. 1972. Diversity and pattern in plants and insects. Ecology, 53: 819–829.
Odum, E. P. 1953. Fundamentals of Ecology. Saunders, Philadelphia.
Paine, R. T. 1969. A note on trophic complexity and community stability. Amer. Nat. 103: 91–93.
Paloheimo, J. E. 1971. On a theory of search. Biometrica, 58: 61–75.
Patten, B. C. 1963. Plankton: optimum diversity structure of a summer community. Science, 170: 894–898.
Pearson, N. E. 1974. Optimal Foraging theory. Quant. Sci. Paper No. 39, Univ. of Washington, Seattle, Washington.
Pimentel, D. 1961. Species diversity and insect population outbreaks. Ann. Ent. Soc. Amer., 54: 76–86.
Pulliam, H. R. 1974. On the theory of optimal diets. Amer. Nat. 108: 59–74.
Rand, A. S. 1967. Predator-prey interactions and the evolution of aspect diversity. Atas do Simposio sôbre a Biota Amazonica, Rio de J., 5: 73–83.

Ricklefs, R. E. and K. O'Rourke. 1975. Aspect diversity in moths: A temperate-tropical comparison. Evolution In press.
Roff, D. A. 1974a. Spatial heterogeneity and the persistence of populations. Oecologia, 15: 245–258.
Roff, D. A. 1974b. The analysis of a population model demonstrating the importance of dispersal in a heterogeneous environment. Oecologia, 15: 259–275.
Rosenzweig, M. L. 1973. Exploitation in three trophic levels. Amer. Nat., 107: 275–294.
Rosenzweig, M. L. and R. H. MacArthur. 1963. Graphical representation and stability condition of predator-prey interactions. Amer. Nat., 97: 209–223.
Schoener, T. W. 1971. Theory of feeding strategies. Ann. Rev. Ecol. Syst., 2: 369–404.
Smith, F. E. 1972. Spatial heterogeneity, stability and diversity in ecosystems. Trans. Conn. Acad. Arts Sci., 44: 309–335.
Tahvanainen, J. O. and R. B. Root. 1972. The influence of vegetational diversity on the population ecology of a specialized herbivore *Phylotreta cruciferae*. Oecologia, 10: 321–346.
Volterra, V. 1937. Principe de biologie mathematique. Acta Biotheoretica, 3: 1–36.
Watt, A. S. 1947. Pattern and process in the plant community. J. Ecology, 35: 1–22.
Watt, K. E. F. 1964. Comments on fluctuations of animal populations and measures of community stability. Canad. Entomol, 96: 1434–1442.

Author's address:

Gordon H. Orians
Department of Zoology
University of Washington
Seattle, Washington 98195
U.S.A.

Diversity, stability and maturity in natural ecosystems

Ramón Margalef

Introduction

Diversity, stability and maturity, like niche and competitive exclusion, have been at the centre of much discussion. By covering a wide spectrum of meanings these terms have been useful in stimulating—and also in muddling—thinking. It now seems that most of the discussions on relationships between stability and diversity lead nowhere, particularly because it is difficult, perhaps impossible, to find direct causal relationships between them, and to explain stability in terms of diversity or vice-versa. This is because both concepts, as they are used, refer to external or peripheral properties of the ecosystem, and any empirical relation between them is a consequence of their common dependence on more fundamental properties of the ecosystem.

In addition to stability and diversity many other ecological terms are also inexact. The biomass in trees, for example, includes much dead tissue, and ecologically such wood is not different from highways, houses, and artifacts used for communication by man. Another widely used concept, that of energy flow, is equally dubious. This covers the energy that is fixed by photosynthesis, but primary production is enhanced by the degradation of energy in other channels or systems (such as evapotranspiration, mixing of water etc) so that it is really impossible to estimate the total amount of degraded energy involved in ecosystem dynamics. All this puts ecology on very shaky foundations, and certifies it as a soft science. But the situation is no better with regard to evolution and other branches of biological sciences.

One source of difficulty is the use of mathematical and statistical models assuming symetrical relationships and a reversible logic. Predictions based on such models often do not work. Many difficulties plague the models relating numbers of species to available resources, describing competition, or the results of the development of defence mechanisms, or the properties of interconnected webs of interspecific relationships, or the probabilities of extinction. It is significant that all such models predict simplification: extinction for some of the competing species, instability in a multiconnected system. Nevertheless, defence mechanisms and competition have been agents of diversification in the biosphere, and not of simplification. A complex system has many possible connections, but is not necessarily unstable; it simply has more possible steady states, and as only one is realized at a time, this one can be selected from among a larger set of available states and most likely among those that can last longer. Any theories that predict to the contrary, that is, a trend towards simplification or instability, must contain something basically wrong.

Nor have ecological models been very successful in other aspects of prediction (except perhaps for some simple interpolation). This is probably inevitable considering the nature of the problems posed by ecology, comparable to those posed by the study of evolution. Explanations have been produced *post facto* and on an *ad hoc* basis. In criticising evolutionary theory it has been argued that biologists are equally capable of producing an explanatory hypothesis or model for the presence of horns in a species of mammal or for their absence, and such situations can be easily ridiculed by practitioners of more 'hard' sciences. We must recognize that the same, or worse, happens in ecology. It is usually argued that the difficulties lie in the complexity and diversity of ecological systems, and it is curious that modest predictions on a more general level based on thermodynamics or similar laws are rarely attempted.

I am convinced that diversity cannot be reduced to a statistical problem, and stability is not a matter of definition, or of multiplying definitions, but it seems to me that these terms, together with maturity, reflect crude impressions about the apparent behaviour of *physical* systems, and I stress physical, meaning that due consideration must be given to the laws of thermodynamics, of natural selection, and of properties that appear in material systems but cannot easily be deduced from the simple and usual mathematical models.

Shortcomings of mathematical models

The difficulty in the construction of sensible models following the usual approaches can be illustrated by the way in which equations of the Volterra-Lotka type are used. It seems to me that the use of models made of sets of differential equations is in much need of a complementary approach, taking into account the study of the behaviour of the system as a whole. Here the connections between the laws of thermodynamics and natural selection must find some expression, and only in such a framework can the problems of organization and strategy be discussed, problems which could be deemed mental artifacts or non-existent in a purely mathematical framework.

The following discussion will serve to present an example of the kind of questions involved, and is essential to my understanding of stability. Stability may have many meanings, but most of them can be connected to some dynamic process in which a manifestation of natural selection can be discovered.

The usual form of multipopulation models is a set of differential equations

$$dN_i/dt = f(A, B, C, \ldots, N_i, N_j, \ldots)$$

frequently written as

$$dN_i/dt = EN_i + \sum a_{ij}N_iN_j$$

Difficulties arise at once because relationships are never as constant as assumed, nor are they necessarily restricted to interactions between quantifiable elements of the system. But we can consider that some of them (a_{ij}, $a_{ij}N_j$, or $a_{ij}N_iN_j$) are random functions, or functions that cannot be studied properly from inside the system. Then we might write

$$dN_i/dt = EN_i + \sum a_{ij}N_iN_j + \sum R_{ik}N_iN_j + \sum R_{il}N_j + \sum R_{im}$$

in which R stands for the values of different random functions.

Every persistent system has by definition achieved some measure of conservatism. Persistent populations manage to keep within narrow limits a sum of terms which includes a certain number of random terms. In this context 'random' means that the values of the variables are not determined within the system under consideration, though they may be determined within the framework of a larger system in which our particular system is embedded. In a given system some relationships are inevitable as they are the properties of the participating organisms, others are optional, and evolution can play the game of combining the options. The results may be consolidated and function as fixed links in future generations. What persists can be associated with or accepted as the result of a strategy of life and the workings of natural selection, and results presumably from the combination, for every species, of different segments of different random functions. Events that are unpredictable over a short period obey a probability law if taken over years. Events unpredictable on a square centimeter scale, may become predictable over the hectare. Food is largely assured if different foodstuffs are available, or if the behaviour of the prey fits the behaviour patterns of the predator in some way.

Stabilization as a result of combining random inputs

As discussed later, the strategy of life is to construct bridges combining events and resources across time and space, or else to anticipate change. I believe it adequate to call stabilization the result of the combination of segments of random functions; in fact the qualification of stable can be given to any chosen combination which persists. Stabilization results from integrating values over time (longer life, accumulation of reserves), over space (mobility, transport), or by combining dependence on factors and resources that were not already linked. Further developments along these lines can, I think, put a merciful end to discussions as to whether population regulation is, or is not, density-dependent, as the strategy can develop in alternative directions according to the circumstances. A diagramatic representation of possible choices is shown in Figure 1. This indicates the generation of stability combining random functions on which survival of the species comes to depend. The alternative to sending long 'roots' across space and time is to keep a high rate of multiplication to compensate for high and unpredictable risks of destruction. This is the only option open in strongly fluctuating climates, in exploited ecosystems, turbulent water and mowed lawns.

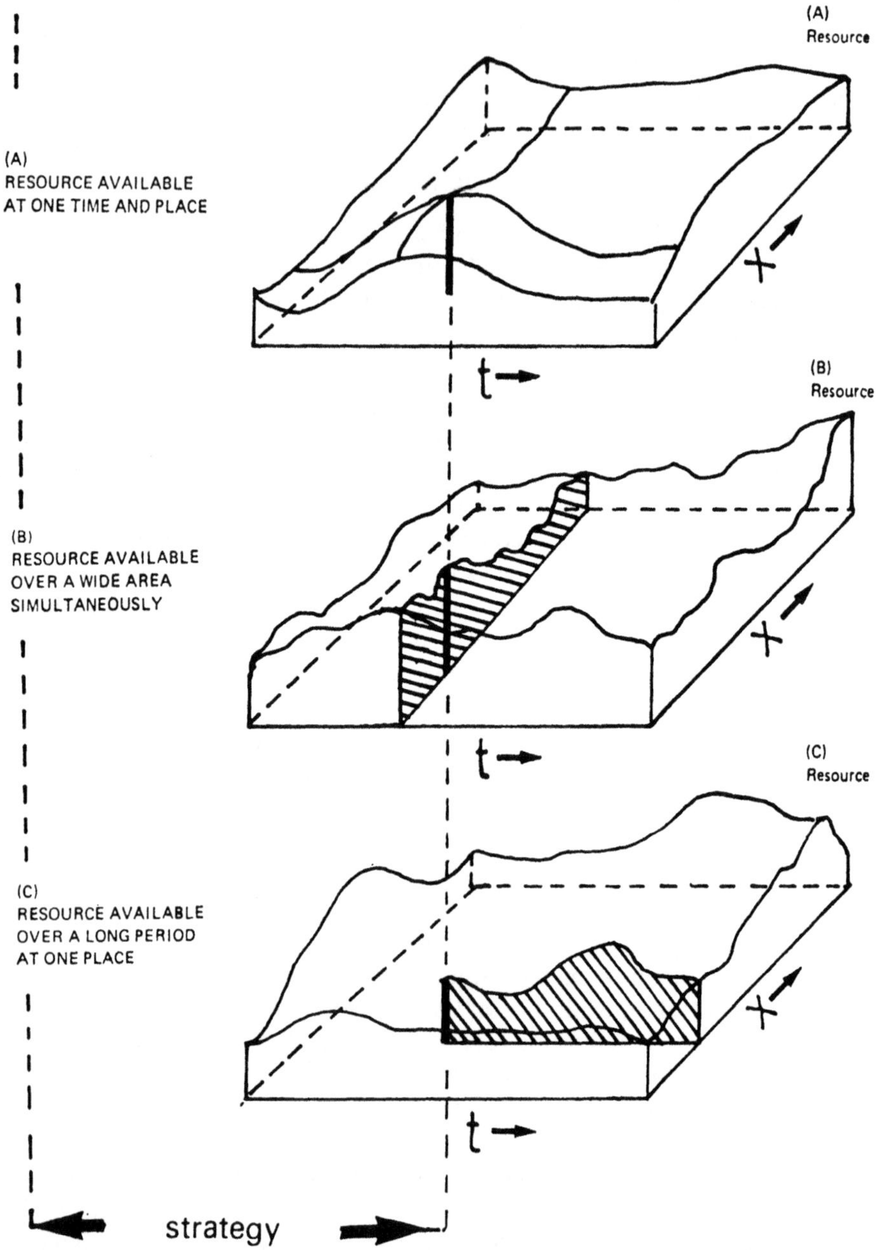

Figure 1. Diagrammatic representation of three variables (*A*, *B*, *C*) in space (*X*) and time (t) representing resources sampled in different ways by one species. Some resources can be obtained only at one time and place (*A*), but several of such resources can be combined; other resources can be obtained over a wide area explored in a short period of time (*B*), or else over an extended period (establishing reserves inside or outside the body) at a particular place (*C*).

In such unpredictable situations, systems are kept rather simple, building up again and again at a relatively high thermodynamic cost per unit of supported biomass. My impression is that even in such circumstances genetic mutation in the species, or the introduction of genotypes from outside, represents a continuous process of trying, failing or succeeding, a process of natural selection, which appears to aim at reducing the range of variation of numbers of individuals in the population, only achieved by an increase in interrelationships in time and space. In the context of any set of differential equations describing kinetics of the different species this means a tendency to narrow the range of variation not only of dN/dt but also of d^2N/dt^2.

The essence of 'maturity'

These considerations also offer the opportunity to clarify the meaning of maturity as I have used this word. Maturity is a non-committal term based on succession, and on the reciprocal probabilities of transition between pairs of systems or states. The transition is more likely to occur from a less mature system to a more mature one than the reverse, and maturity is associated with an increase in internal effective links (development of homeostasis) by becoming less dependent on inputs not determined within the same system. The whole process of trial and error and stabilization conducted by species within an ecosystem proceeds along an axis of increasing maturity. At low maturity a system remains highly reactive and has to cope continually with new and unpredictable inputs. This may result in the acquisition of information that can sometimes be used later (as, for instance, the development of rhythmic behaviour). In a mature system, a larger proportion of the incoming inputs have been anticipated, or they can be ignored because an appropiate reaction to them is no longer essential for survival. Survival and stability of populations have been achieved through selection of appropiate factors; the intensity of each one of these may change, but their total effect remains predictable by a gradual selection of them, and this means spreading the risk in the game of existence. Life is always processing information that is used to block further inputs of information; when survival is assured by certain means, many sources of information not related to these means become irrelevant.

The role of natural selection

It is true that the number of possible situations may be almost infinite and that statements of a general nature made in ecology appear wrong or arbitrary since it is almost always possible to produce some evidence to the contrary. The dictum that only the survivors write the history can be applied to the biosphere, and we boast that a number of strategies may be recognized among the species that survive. Many strategies within an ecosystem appear to be complementary; for instance, a predator may be a kind of K-strategist (MacArthur and Wilson, 1967), primarily adapted to biotic elements, while its prey may have to remain closer to the *r*-strategy, geared primarily to abiotic factors. At every part of the earth's surface there is a possibility of destruction of the biosphere, making space for the colonization and survival of pioneer species and ecosystems. But, as has been assumed before, it seems that in the evolution of species there is a higher probability of passing from *r*- to *K*-strategy than the reverse, and in ecological succession a stabilizing factor for the ecosystems consists of the increasing importance of species that establish connections more and more extended over time and space. In practice this amounts to *K*-strategists taking over the control of the ecosystem. Moreover, I consider that natural selection at the level of the species, and acceptance or rejection of potential new members of the community in ecological succession, are both bound by laws of economy, in the sense that energy flow per unit biomass tends to be kept as low as circumstances and history allow.

Following such ideas, it is possible to complement the sets of differential equations typical for the usual ecological models, and to which reference has already been made. Further development of the reasoning involves consideration of a set of natural laws to which both the laws of thermodynamics and natural selection belong; this reasoning is difficult, but the basic evidence is in every system and easy to grasp, and can be illustrated briefly here by reference to two types of ecosystems.

The slowing down of turn-over during succession

Consider a column of water, for the moment well stratified, and consider the development of life within it (Fig. 2). The probability of nutrients moving down the column is higher when they are in particulate form or in the bodies of organisms, than when they are in solution. Gravity and animal migration displace the nutrients downward until a steep gradient of nutrients in solution is formed, limited by physical diffusion. If a parcel of deep water rich in dissolved nutrients comes up to the well-lit zone, primary production is increased and immediately mechanisms appear that lead to the recovery of the equilibrium situation. It looks as though in the organization of ecosystems life is always slowing down its own activity. This can be expressed as a decrease in the primary production/biomass ratio from initial colonization of a vacant water column, along a process of succession, to end in a community with a well-differentiated vertical organization controlled mainly by the activity of animals. The existence of cybernetic (feedback) mechanisms is implied, and consequently some reference value has to be assigned to the variables. This reference point, in effect, represents the extrapolation of the workings of the ecosystem, and in the present case may be the one compatible with a minimum value for the primary production/biomass ratio. Departure from this situation immediately activates mechanisms that return the system to the reference point. This property can be considered essential to the stability or elasticity of the system and is determined by the strategies of the species that appear to assure the workings of such mechanisms. It is,

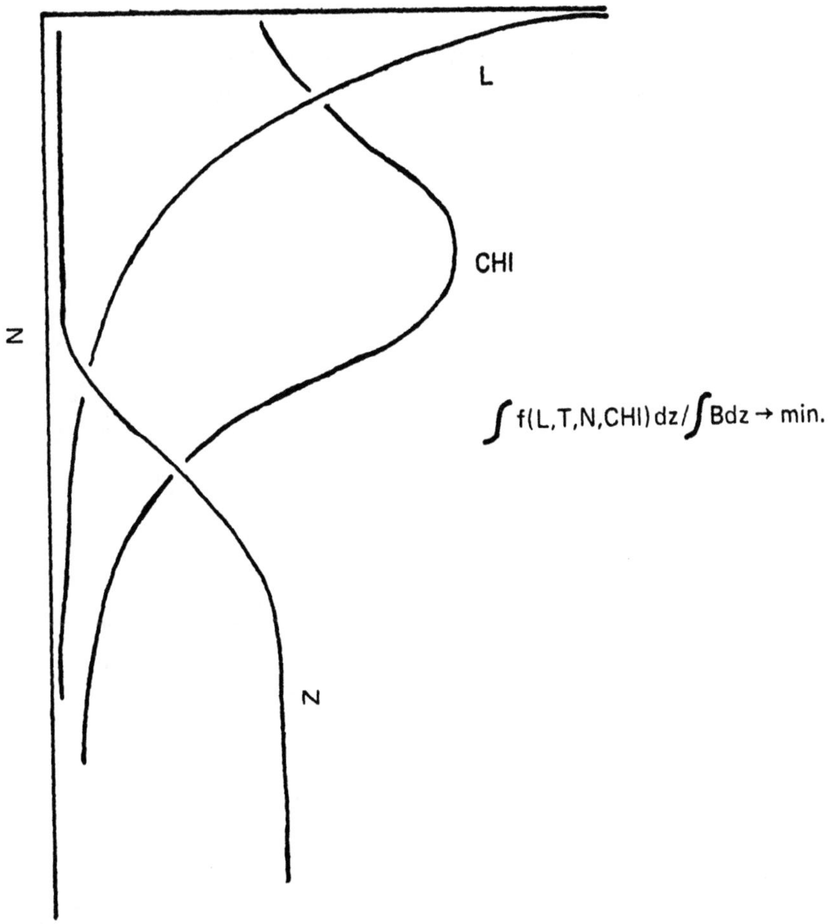

Figure 2. Vertical distribution (Z-axis) of light (L), nutrients (N) and chlorophyll (Chl) in a water mass, and tentative expression of a general law of development of the ecosystem. (T = temperature; B = biomass.)

perhaps inevitable, and probably harmless, to use a finalistic terminology in this context, although it is clear that the mechanisms involved (oxygen valve, loops in the cycles of phosphorus, carbon, nitrogen, oxygen, passing outside the system, in the sediment or in the atmosphere) have nothing mysterious about them.

A terrestial system works the same way (Fig. 3). In a forest we have a system in which competition for light and defence against animals have created that piece of magnificient vertical organization and transport called a tree. But growth in height immobilizes increasing amounts of organic matter and lengthens transport time, slowing down turnover. As in water, the development of the system splits the biosphere into adjacent structures each with a well-organized vertical transport system, which in practice interrupts horizontal

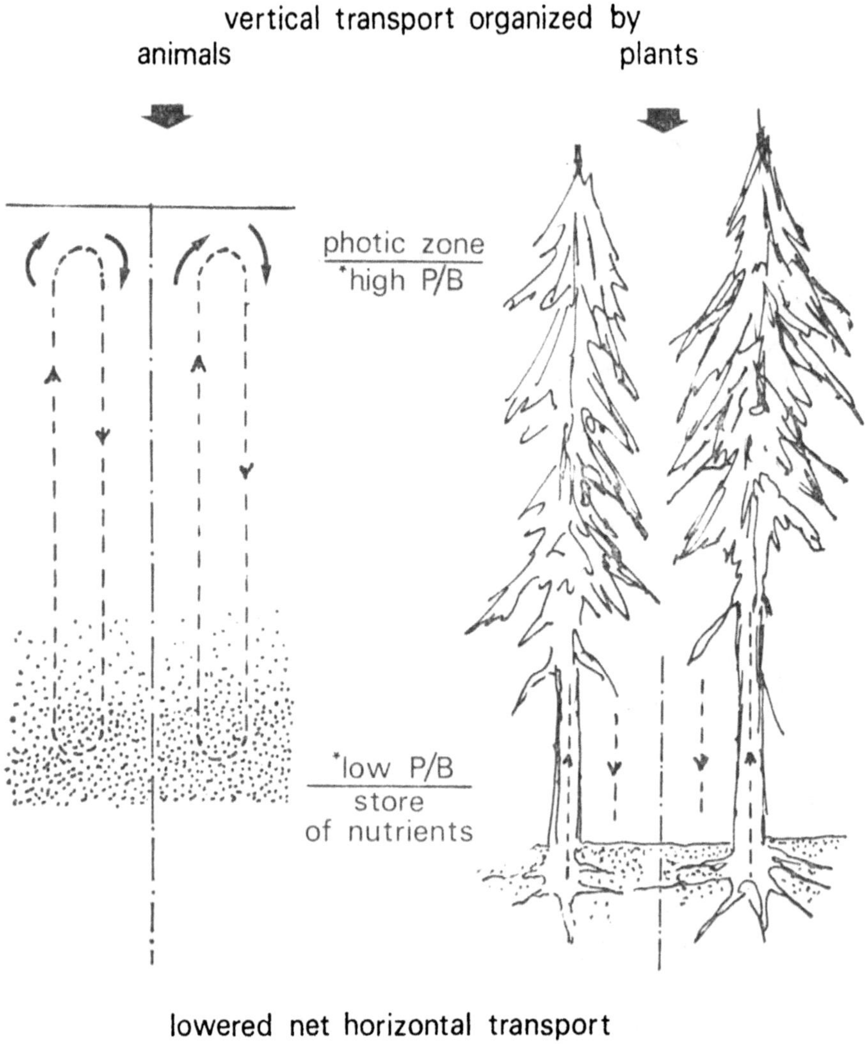

Figure 3. The analogous development of transport systems in aquatic and terrestrial ecosystems: vertical organization develops with competition for light, the action of gravity, and the activities of animals. (P/B = production/biomass ratio.)

transport between neighbouring structures. The transport system is, however, controlled mainly by plants in terrestial ecosystems (apart from interference by man), and mostly by animals in aquatic ones. From this result the most important differences between aquatic and terrestrial ecosystems, expressed in their plant biomass/animal biomass ratio, and in their exploitability by man.

The trend to minimize the ratio between primary production and biomass can be accepted for any closed ecosystem, that is, if it receives only radiation from outside. The recognition of such a trend, a result of inherent mechanisms in the ecosystem, probably needs the acknowledgement of other associated properties working simultaneously, for instance that changes occur in such a way as also to minimize the time necessary to attain the final state. This means simply that in the course of succession fast-growing species are the first to develop, and these are gradually substituted by species with more interrelationships, able to compete successfully and having a slower turnover.

The coupling of ecosystems

In nature all this cannot be more than an approximation since there are no completely closed systems. In the real systems, in addition to received radiation used in photosynthesis, some extra energy is effective in promoting mixing, changes in density of thermal origin, circulation, evapotranspiration etc. Consequently, our initial system to which reference is made appears to be coupled to another, and this coupling is likely to be more important if there is some exchange between them in the form of nutrients or of organisms. It is then necessary to reconsider the prediction of a decreasing turnover rate and to reassess it for the whole or compound system, for the exchange between coupled systems may involve a reactivation or acceleration of turnover in the small subsystem. For such coupled systems we tend to say that any component system is made unstable, that it is not self-supporting and therefore cannot be understood without also considering the larger system in which it is embedded. The same reasoning may be applied to any combination of ecosystems. Cropped land may seem unstable, and in process of stabilization, because it returns to a state of lower production/biomass ratio when it is not exploited; but it can be seen as stable when coupled with an exploiter (man) and made part of a larger system that then, of course, includes man.

Conclusion

I think that all these processes: the strategy of single species during evolution, their replacement in the course of succession, and the action of many natural mechanisms that slow down the turnover in ecosystems, can be encompassed in a general view of ecosystem dynamics. There is a continuous shift in the conditions of selection and the possibilities for stabilization that cause substitution of some species by others, and with the passage of time clusters of coevolving species appear. In speaking of selection at the level of the ecosystem nothing more is implied. The aspect that I want to stress again is that in the infinite number of possible dynamic concordances between environments and communities, there is always something that tips the balance—or may be the inertia of nature—towards making relationships more complex and predictable and slowing down turnover. As the fulfilment of such a trend means a smaller amount of energy exchanged per unit of biomass maintained, there are reasons to seek links with the laws of thermodynamics. Here ecological stability may have a quasi-physical meaning. But it is perhaps questionable whether the term stability should be retained, as it has been used too much in different and divergent speculation.

Criticism of diversity does not perhaps lead so far. I stress that diversity has nothing to do with simple statistics, but with the expression of the dynamic properties of a complex system. Organization consists of hierarchical couplings between subsystems of necessarily different turnover. This kind of structure places constraints on any subdivisions of sets into subsets, whatever the adopted criteria of classification. I consider that diversity is any useful measure of how a set (system) falls into subsets (subsystems), and is important as a measure of the intensity of interactions taking place. Consider a chemical system in which the proportions of the different reagents are defined by a set of equilibrium constants. A diversity can be computed from the concentration of the different chemical substances present. The addition of one reagent, or separation of a product, alters the ratios and thereby changes diversity. It is easy to understand why diversity drops in an ecosystem that undergoes transition, or becomes an open one, as happens when one system is coupled with another (Margalef, 1974, p. 368).

Measures of diversity refer to a point in time, and stability refers to a sequence of states. If the individual is the unit, the species (the subset) a team in the game, and the whole community (the set) represents all the players engaged in the game, then both diversity and stability have a demographic and taxonomic interpretation. It seems to be a common occurrence that a complex of circumstances allowing a high diversity also permits a high stability or constancy in taxonomic composition: nature tends to become baroque in situations permitting a high maturity, with little energy left for large changes. If interest is focussed on other properties, such as energy flow, the picture may look different. A system which is highly unstable in species composition may be stable with relation to the energy flowing through it.

References

MacArthur, R. H. and E. O. Wilson. 1967. The theory of island biogeography. Princeton Univ. Mon. Pop. Biol., 1: 1–203.

Margalef, R. 1974. Ecologia. Ediciones Omega, S. A., Barcelona, 951 pp.

Author's address:

Ramón Margalef
Department of Ecology
University of Barcelona
Barcelona
Spain

Stability in ecosystems: some comments

Robert M. May

Introduction

The variety of meanings that can be attached to 'stability' in ecological contexts has been discussed by Lewontin (1969), Holling (1973), May (1973) and others. Orians (1975) has opened this section with an insightful review of this material.

My paper is restricted to a few disconnected comments, made with a view to sharpening discussion. Some of these (in the first section) pertain to various technical aspects of model ecosystems, and others (in the second section) to biological morals which emerge from such studies.

General mathematical characteristics of model ecosystems

Structural stability

Commonsense requires that models in population biology be structurally stable, in the sense that their predictions are not qualitatively dependent on the details of the mathematical functions used to describe the interactions between species. This is very different from physics, where the analysis often hinges upon special (structurally unstable) symmetries and conservation laws.

For instance, many properties of the periodic table follow from the inverse square, R^{-2}, Coulomb force law; an $R^{-1.5}$ law would give quite different properties. But our ecological understanding is such that we can only be interested in conclusions which are, as it were, valid for all force laws from $R^{-1.5}$ to $R^{-2.5}$.

Non-linear versus linear equations

The equations describing model ecosystems are non-linear, and as such can exhibit a much wider range of dynamical behaviour than can linear systems.

Thus, in non-linear equations, stable limit cycle solutions (wherein populations oscillate stably between maximum and minimum values determined by the parameters of the equation) are as pervasive and natural as are the stable equilibrium points so familiar from elementary mathematics courses on linear systems. Examples (May, 1973) of such stable cycles are to be found in: (1) predator-prey systems; (2) continuously growing single populations with explicit time delays in their regulatory mechanisms; (3) single populations with non-overlapping generations (i.e. systems with discrete growth, described by difference equations: the relation between such models and those with explicit time-delays is discussed by May et al., 1974); (4) systems with three or more competitors; and others.

Moreover, even the simplest non-linear difference equations can manifest further complications. As the intrinsic growth rate, r, increases, the stable equilibrium points give way to a hierarchy of well-defined stable cycles (of period 2^n), which at yet larger r give way to a regime of chaotic behaviour, where the system can show cycles of arbitrary period, or even totally aperiodic behaviour, depending on the initial population values (Li & Yorke, 1975; May, 1974; May & Oster, 1975). That such a spectrum of dynamical behaviour is latent in simple and deterministic difference equations is a fact which deserves to be more widely appreciated. It has disturbing implications.

Stability, complexity and environmental predictability

We now turn to see what biological messages emerge from the study of general mathematical models for ecosystems.

Some formal discussion is first necessary.

Suppose we have a community of m interacting species, with the population of the ith species at time t represented by $N_i(t)$. A mathematical model of this system will aim to describe the rates at which the populations change, dN_i/dt, dependent upon the values of various interaction and environmental parameters, κ_j (such as birth rates, assimilation efficiencies, predator attack rates, etc), and upon the magnitudes of the populations, $N_i(t)$, themselves.

In assessing the stability character of any such model, various factors enter. First, equilibrium configurations will exist only within certain restricted ranges of the interaction and environmental parameters, κ_j; outside this restricted region of parameter space (κ-space), the equations will describe a collapsing ecosystem in which some, or all, of the m populations are fated for extinction. Mapping out this stable region of parameter space is a standard exercise in physics and engineering problems. Second, even when the parameters are fixed at values which do admit of an equilibrium solution, this equilibrium may be unstable to large perturbations in the populations (this is, in Orians', 1975, terms, the question of 'amplitude stability').

These two questions are intertwined, both mathematically and biologically. For parameter values near the centre of the domain of stable values, the dynamical landscape in population space may in general be thought of as a comparatively wide and deep valley; all but the most extreme disturbances to the populations will eventually return to the equilibrium configuration. As the parameter values are chosen nearer the edge of the domain of stable values, the dynamical landscape in population space becomes more like a shallow valley nestled in the top of a volcano; modest disturbances to the populations will see the system spill out from the volcano top. Finally, for parameter values outside the stable domain, there is no valley at all; the volcano tip has given way to a rounded hilltop. Moreover, in the real world, we do not deal with fixed parameter values, but rather the environmental and interaction parameters are themselves fluctuating, in turn driving the population perturbations.

In short, a system which is stable only within a comparatively small domain of parameter space (schematically illustrated by Fig. 1b) may be called *dynamically fragile*. Such a system will persist only for tightly circumscribed values of the environmental parameters, and will tend to collapse under significant perturbations either to environmental parameters or to population values. Conversely, a system which is stable within a comparatively large domain of parameter space (schematically illustrated by Fig. 1a) may be called *dynamically robust*.

With these preliminaries disposed of, a useful result can be stated.

A wide variety of mathematical models suggest that as a system becomes more complex, in the sense of more species and a more rich structure of interdependence, it becomes more dynamically fragile. Models exhibiting this effect range from specific detailed ones, through more abstract ones described by community matrices, to models which describe only the topology or 'loop structure' of the food web (May, 1973; Levins, 1974). Figure 1 aims to illustrate this: Figure 1a depicts the stable domain of parameter space for a relatively simple community,

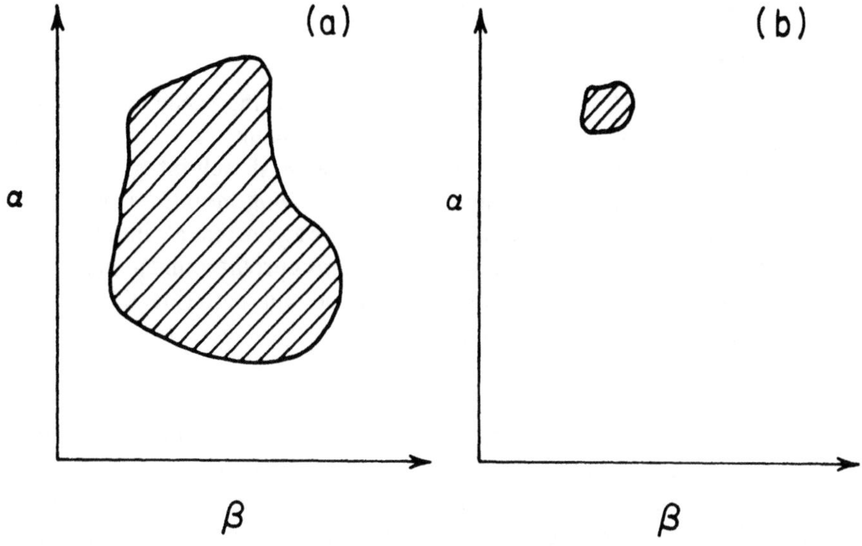

Figure 1. (a) Schematic representation of a two-dimensional cross-section through the stable region of parameter space for a simple system (few species, few interaction parameters). (b) Likewise, for a complex system (many species, and a parameter space of many dimensions).

which will tend to be dynamically robust; Figure 1b is for a relatively complex community, which will typically be more dynamically fragile.

This is not to say that, in nature, complex systems need appear less stable than simple ones. Suppose the complex system of Figure 1b is in an environment characterized by a low level of random fluctuations, whereas the simple system of Figure 1a is confronted by a high level of random environmental fluctuation: the two systems can well be equally likely to persist, each having the dynamical stability properties appropriate to its environment.

Using Slobodkin's (1964) image of evolution as an existential game, where the prize to the winner is to stay in the game, we may conjecture that ecosystems will evolve to be as rich and complex as is compatible with the persistence of most populations. In a predictable environment, the system need only cope with relatively small perturbations, and can therefore achieve this fragile complexity, yet persist. Conversely, in an unpredictable environment, there is need for the stable region of parameter space to be extensive, with the implication that the system must be relatively simple.

In brief, a predictable ('stable') environment may permit a relatively complex and delicately balanced ecosystem to exist; an unpredictable ('unstable') environment is more likely to demand a structurally simple, robust ecosystem.

This idea has several implications, some of which will be developed immediately below. The picture is, of course, vastly oversimplified: a few among the many qualifications and complications, which are necessary in any detailed picture, are touched on at the end (see also Whittaker, 1975).

Complex natural ecosystems are fragile

One corollary is that, in this scheme, the large and unprecedented perturbations imposed by man are likely to be more traumatic for complex natural systems than for simple ones. This inverts the naive, if well-intentioned, view that 'complexity begets stability', and its accompanying moral that we should preserve, or even create, complex systems as buffers against man's importunities. I would argue that the complex natural ecosystems currently under siege in the tropics and subtropics are less able to withstand our battering than are the relatively simple temperate and boreal systems.

Natural monocultures are often very stable

On the above view, there is no reason to expect simple natural monocultures to be unstable. And, indeed, there are many instances of robustly enduring natural monocultures. The marsh grass *Spartina* is one conspicous example. Tischler (1972) has emphasized other examples, mainly in littoral systems, where the stable natural system is essentially a monoculture. Another currently relevant case is bracken (Lawton, 1974). In recent years, partly as a result of hilly areas grazed by cows being given over to sheep, bracken has shown itself to be a robust, even aggresively invasive, natural monoculture over increasing areas in Britain.

The instability of so many man-made agricultural monocultures is likely to stem not from their simplicity, as such, but rather from their lack of any significant history of coevolution with pests and pathogens.

Mutualism

Models for obligate mutualism tend to be dynamically fragile: such systems are liable to extinction if one or other partner fluctuates to low population densities. This may help explain why mutualistic interactions feature more prominently in relatively predictable environments.

Quantifying 'resilience'

These ideas, and particularly Holling's (1973) penetrating discussion of the concept of resilience, prompt a practical question. How, in natural systems, can we measure the size of the stable domain (the size of the hatched areas in Fig. 1)? How quantify dynamical fragility or robustness? Ecological engineering would come of age if the theoretician could satisfy the empiricist's need for a well-defined recipe: a list of experiments and measurements to make, a data array to calculate, some quintessential number to be distilled.

I find it difficult to envision any simple number, or handful of numbers, which will quantify the resilience of a complicated natural ecosystem. In real systems of any degree of complexity it seems likely that the capacity to withstand perturbations will depend on the kind of perturbation; that the hypervolumes hinted at by Figure 1 will have funny shapes; that the valleys nestled in the volcano tips may have sides which are short but steep in one direction, far extended but shallow in another. Such systems are liable to be resilient under large natural perturbations (e.g. violent tropical storms) with which they coevolved, but vulnerable to apparently minor disturbances of a novel kind (e.g. bisection by a road or electric power line).

One moral I draw is that, at least in the near future, success in confronting theoretical ideas about stable domains with real world observations will come from the detailed study of simple, low-dimensional systems or subsystems. Halting first steps in this direction are represented, for example, by the work of Hassell and others (Hassell & May, 1973; Hassell, 1975; Beddington et al., 1976) on insect host-parasite systems. This academic observation is not much help to people facing pressing environmental problems, where all too often sloppy 'environmental impact' studies do not even include simple species counts.

As mentioned previously, the above discussion of the relation between biotic complexity and environmental predictability is grossly oversimplified. Some additional remarks, which make my presentation more muddy but more honest, are as follows.

Environmental unpredictability tends to diversify reproductive strategies

We have argued that a relatively predictable environment will permit the evolution of a complex and fragile system. This predictability is, furthermore,

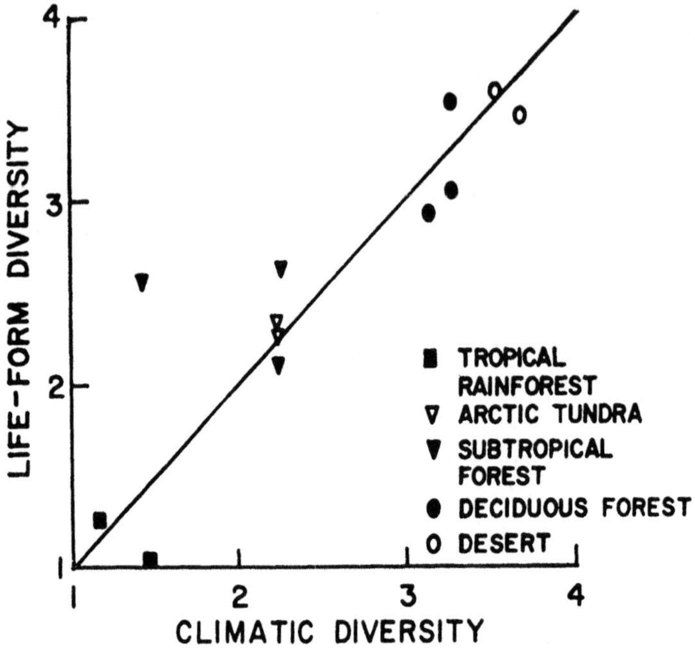

Figure 2. Illustrating the positive correlation between rainfall unpredictability ('climatic diversity') and diversity of reproductive strategies, for various plant communities around the world. From Givnish (1975), where details are given.

likely to mean that certain basic themes will be selected as optimal, and then specialized variations rung upon these themes. Thus the myriad of tropical trees and plants essentially all opt for the same basic strategy for the placement of their reproductive parts (viz., phanerophytes). Conversely, a highly unpredictable environment will require a structurally simple system, without opportunity for such thematic elaboration; but on the other hand this unpredictability can favour the evolution of many alternate strategies for reproduction or annual regeneration (i.e., the evolution of many alternate Raunkiaerian life- forms). The observation that environmental unpredictability can lead to diversity in reproductive strategies has been made by Whittaker (1974, 1975) and others, and has been graphically documented by Givnish (1975): see Figure 2. This figure plots the diversity of reproductive strategies (as characterized by the relative abundance of the different Raunkiaerian life forms) against a measure of the rainfall unpredictability, for a variety of plant communities from tropics to arctic: for details, see Givnish (1975).

Natural disturbances and succession as a source of diversity

Succession, which in Whittaker's words is 'an ecotone is time', can also complicate the discussion. It has been pointed out by several people (notably by Connell (1975) for rainforests subject to tropical storms, and by Levin and

Paine (1974) for the rocky intertidal subject to battering by logs) that openings are continually being created even in stable ecosystems, and that the consequent simultaneous presence of all successional stages is a significant contribution to the species richness of the system, viewed in the large. Describing the successional process by a Markovian transition matrix, Horn (1974, 1975) has discussed the properties of the overall steady-state mix of early and late successional species. This work should be read. Very briefly, it suggests that some finite degree of disturbance (back to the zeroth successional state) is necessary if the system is to realize its potential diversity, but that a high frequency of disturbance will in general lead to a system of relatively low diversity.

Morals

Amongst this potpourri of comments, three general points deserve emphasis.

The simplest models of ecological systems are necessarily non-linear, and as such their equilibrium solutions can be constant, or cyclic, or can even be chaotic but bounded fluctuations.

Complex ecosystems (such as the tropical rainforest or Lake Baikal, with their many species and rich interaction structure) are in general dynamically fragile. Although well adapted to persist in the relatively predictable environment in which they have evolved, they are likely to be far less resistant to the perturbations imposed by man than are relatively simple and robust temperate ecosystems.

Natural monocultures are often highly stable (e.g. *Spartina*, bracken). The thing which destabilizes man's agricultural monocultures is not so much their simplicity per se, as their lack of an evolutionary pedigree.

References

Beddington, J. R., M. P. Hassell and J. H. Lawton. 1976. The components of arthropod predation. II The predator rate of increase. J. Anim. Ecol. 45 (in press).

Connell, J. H. 1975. In: Ecology of Species and Communities. J. M. Diamond & M. Cody eds. Harvard Univ. Press, Cambridge, Mass. (To be published).

Givnish, T. 1975. The contribution of seasonality to plant diversity. (To be published).

Hassell, M. P. and R. M. May. 1973. Stability in insect host parasite models. J. Anim. Ecol. 42: 693–726.

Hassell, M. P. 1975. Density dependence in single species populations. J. Anim. Ecol. 44: 283–295.

Hassell, M. P., J. H. Lawton and J. R. Beddington. 1976. The components of arthropod predation. I The prey death rate. J. Anim. Ecol. 45 (in press).

Holling, C. S. 1973. Resilience and stability of ecological systems. Ann. Rev. Ecol. Syst. 4: 1–23.

Horn, H. S. 1974. The ecology of secondary succession. Ann. Rev. Ecol. Syst. 5: 25–37.

Horn, H. S. 1975. Marokvian properties of forest succession. In: Ecology of Species and Communities. J. M. Diamond & M. Cody eds. Harvard Univ. Press, Cambridge, Mass. op. cit. (To be published).

Lawton, J. H. 1974. The structure of the arthropod community on bracken (*Pteridium aquilinum* (L.) Kuhn). In: The Biology of Bracken. F. H. Perring ed. Academic Press, New York.

Levin, S. A. and R. T. Paine. 1974. Disturbance, patch formation and community structure. Proc. Nat. Acad. Sci. U.S.A. 71: 2744–2747.
Levins, R. 1974. The qualitative analysis of partially specified systems. Ann. N.Y. Acad. Sci., 231: 123–138.
Lewontin, R. C. 1969. The meaning of stability. In: Diversity and Stability in Ecological Systems. Symp. Biol. Brookhaven Nat. Lab., Springfield, Va. 22: 13–24.
Li, T-Y. and J. A. Yorke. 1975. Period three implies chaos. American Mathematical Monthly (in press).
May, R. M. 1973. Stability and Complexity in Model Ecosystems. Princeton Univ. Press, Princeton.
May, R. M. 1974. Biological populations with non-overlapping generations: stable points, stable cycles and chaos. Science 186: 645–647.
May, R. M., G. R. Conway, M. P. Hassell and T. R. E. Southwood. 1974. Time delays, density dependence, and single species oscillations. J. Anim. Ecol. 43: 747–770.
May, R. M. and G. F. Oster. 1975. Bifurcations and dynamic complexity in simple ecological models. (To be published).
Orians, G. H. 1975. Diversity, stability and maturity in natural ecosystems. In: Unifying Concepts in Ecology. W. H. van Dobben & R. H. Lowe-McConnell eds. Dr. W. Junk, The Hague, Pudoc, Wageningen.
Slobodkin, L. B. 1964. The strategy of evolution. Amer. Sci. 52: 342–357.
Tischler, W. 1972. Discussion remarks at the XIV International Entomological Congress, Canberra.
Whittaker, R. H. 1974. Climax concepts and recognition. In: Handbook of Vegetation Science, Vol. VIII. R. Knapp. ed. Dr. W. Junk, The Hague. 139–154.
Whittaker, R. H. 1975. Stability in plant communities. In: Unifying Concepts in Ecology. W. H. van Dobben & R. H. Lowe-McConnell. eds. Dr. W. Junk, The Hague, Pudoc, Wageningen.

Author's address

Robert M. May
Biology Department
Princeton University
Princeton, N.J. 08540
U.S.A.

The design and stability of plant communities[1]

R. H. Whittaker

Introduction

A striking, and unwelcome, feature of ecology has been the lack of a bridge of logical development and common understanding between population ecology and community ecology. As a synecologist I have been gratified by such recent developments toward this as the work of Robert May. I should like now, in away different from May's and with land plants particularly in mind, to venture some proposals as to how we might connect our understanding of two kinds of living systems—populations and communities.

A stable species pair

Some effective statements on plant populations and communities have been made by Harper (1967, 1969; Harper & White, 1971, 1974). Let me begin, however, with a simplest case—an imaginary forest dominated by two imaginary tree species that I might call, say, antibeech and antimaple. Two key observations about the forest are: (a) its composition is relatively stable, with the two populations existing in a persistent balance of 70 trees of antibeech to 30 of antimaple; and (b) as we move along a continuous environmental gradient—up a mountain, say—the proportions of this balance shift continuously. The two tree species are competitors; they use the same light and shade one another's foliage, and they draw water and nutrients from the same soil. The key question is then how the balance between them is maintained. Suppose we regard the floor of the forest as a mosaic of unit cells, or tesserae, each of which might be occupied by one plant. It does not matter that these are cells without boundaries; let us call these cells microsites. Our two species may each occupy a fraction of the microsites available, while another fraction remain vacant, and while new microsites are being made vacant by the deaths of plants and a fraction of the vacant microsites are being occupied by seedlings. Then, as Skellam (1951) has pointed out and Levin (1974) has developed, in certain cases a stable balance between the populations is possible if environment remains constant, even if the microsites are identical. Two species may form a balance that is neutrally stable if, at

[1] Contribution from research supported by the National Science Foundation. I thank D. Goodman, S. A. Levin and R. B. Root for comments on the manuscript.

the particular point along an environmental gradient that the microsites represent, the species are competitively equivalent, having equal ratios of their death rates to the mean numbers of progeny per parent that are viable after seed predation and will establish themselves if they reach vacant microsites. Possible cases of stable equilibria include: (a) a species with a competitive advantage, versus a species with a reproductive advantage (Skellam, 1951), (b) at least one of the species is stabilized by a constant flow of propagules (that do not reach and occupy all microsites) from a larger area (Levin, 1974), (c) continuing local or patchy disturbance, with the species differing in timing in their use of microsites following disturbance (Levin & Paine, 1974, 1975), (d) patches of initially identical microsites are colonized and held by one or the other of two species which (with dispersal from the patches into other microsites limited) can then coexist (Horn & MacArthur, 1972; Levin, 1974), (e) patches of one species enlarge and break up (by autotoxic or other effects in their centers), and the abandoned microsites are occupied by the other species, and (f) microsites are biologically modified in such a way that seedlings of each species survive better in microsites influenced by large individuals of the other species. The last may be true of antibeech and antimaple. It has been suggested for some forests of beech and maple (F. E. Smith, personal communication) and tropical rain forests (Janzen, 1970; Connell, 1971). Let us term a relative stability of two or more species populations moving in a mosaic of initially identical microsites, in which each species is losing and gaining microsites (while some microsites may remain vacant), a mosaic steady state or mosaic balance. In a real community the microsites will differ, and we modify the mosaic balance accordingly. In some microsites environmental difference that does not result from biological modification implies that competing seedlings of one species or the other are more likely to survive. Other microsites are equally favorable for both species so that the first arrival excludes competitors in these microsites, and a mosaic balance may apply directly to these. I suggest these inferences:

1. In principle mosaic balances and their modification to allow for microhabitat difference may provide interpretation of some of the relatively stable balances we seem to see among plant populations in stable communities.

2. We also observe difference in resource use by adult plants—root depth, for example—and manner and timing of reproduction. We infer that these are niche differences that can, in various combinations with the differences in seedling microsite preference we are assuming, permit continued coexistence of plant species that are all dependent on the same major resources—light, water, nutrients and CO_2.

3. Differences in predation are also likely to be part of the balances. Each species population functions in a steady-state cycle involving density of seeds, frequency of these in suitable microsites, survival of seedlings and growth to reproductive

age as affected by resource use, competition, predation, symbioses and environmental hazard, and density of seeds that the population of surviving, mature plants sheds on the microsite mosaic (Whittaker, 1969; Harper & White, 1971). The relation between available microsites and frequency of seeds in these may have a key role in regulating this cycle, but all the other factors mentioned may affect the flow of plants through the cycle and therefore the frequency of seeds in microsites.

4. We need only shift continuously the relative proportions of the microsites favorable for each species, or the mean characteristics of microsites as affected by climate external to the community, to account for the continuous shift in relative importance of our two species along an environmental gradient that we observed.

A stable community

We need, however, to extend this reasoning to a stable community of many species. It may be possible for three or more species to exist in balance along a microsite gradient. It seems more clearly the case that microsites may differ in a number of characteristics—light, moisture, depth to rock—that can contribute to such balances. Apart from these, the physical and chemical properties of the soil show small-scale variation that may be as significant to the seedlings as it is unwelcome to those who make measurements on soils. We may then conceive of our microsites as ordinated in a hyperspace defined by these physical and chemical microsite factors. Each plant species has its center of preference, or of maximum seedling survival, in this hyperspace; and the species preferences will differ, so that their populations tend to fill the hyperspace. We thus conceive a multispecies and multidimensional modified mosaic balance among plants in microsites. We must consider also the biological influences on the microsites. Each microsite has as factors impinging on it the depth and chemical character of the leaf litter that it receives, the nutrients and allelochemics brought to it by decomposition and rain-wash, and its particular occupation by roots, fungi and soil animals. These factors add to the axes of our hyperspace, and hence the possibilities for difference in plant species responses to and positions in that hyperspace. By this complexity of influences and responses we may at least conceive of the marvellously subtle pattern of undergrowth of an Appalachian cove forest or Mexican oak woodland that we observe, take some superficial measurements of contagion and species-association in, and wish we understood. From this pattern I draw a few more inferences:

1. One notes the self-elaborating quality of the pattern. The more species there are, the more various their influences on the microsites, and the more different microsite qualities are provided to maintain the balances among species populations present and new species that may be added.

2. Communities no doubt differ in the relative significance of the biological, versus the physical and non-biological chemical, influences on microsites. We suppose that the relative significance of the biological influences on microsite difference, and thereby on the maintenance of plant and community species diversity, increases from boreal to tropical forests.

3. This is not to overlook the contribution of symbiosis and predation. We may think of the individual species population in a stable community as being held, contained in an elastic way, in a fraction (a cloud-like portion) of the microsite hyperspace, through which its population flows in a steady-state cycle, contained in this fraction by the characteristics of its own population function and by the pressures on it of competition and predation.

4. The proportion of the microsites occupied, along with the sizes of the plants themselves, affect the share of the community's resources that flow through the metabolism of a species population. Dominance-diversity curves based on productivity tell us something—though not very much (Cohen, 1968; Whittaker, 1969)—about differences among communities and taxocenes in the way resources are divided among species. We note that the population balances to this point involve flows on two levels—the flow of seedlings through the microsite mosaic or hyperspace, and the flow of resources through the metabolism of individuals that supports their growth and reproduction. The community is thus a flow system of multiple component flow systems balanced against one another. The German word well characterizes the kind of stability we are describing so far—*Fliess*gleichgewicht.

Environmental fluctuation

I have so far pretended that environment was constant to construct a conception of the community; but environment is not constant. Let us conceive of two environmental variables—one a spatial variable within the community such as mean moisture conditions of different microsites, and the other a temporal variable such as more or less humid weather that alters the moisture conditions of microsites. We may ordinate the microsites along the spatial variable and plot population numbers along the gradient as in Figure 1 below, assuming an ordinary sort of bell-shaped distribution. The cumulative version of this from left to right, from most to least favorable microsites, is the lower curve. Let the external variable now vary so that some part of this range of microsites becomes untenable and some part of the population dies, along the curve from right to left. If the environment changes from very to not so favorable, affecting individuals in the least favorable sites, range *a*, the effect on the population is relatively small. As the environment changes back and forth in range *b*, marked fluctuation in the population results. As the environment varies in range *c* smaller changes in the population occur as it becomes rare, though persistent in most favorable microsites. In adverse periods those individuals in the most favorable microsites

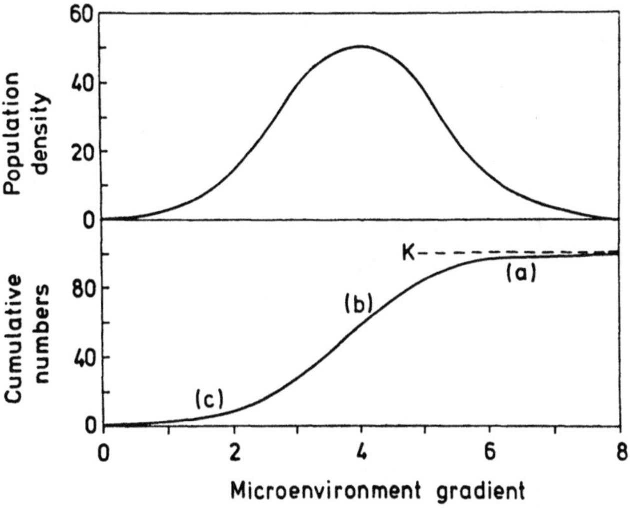

Figure 1. Population response to a microenvironmental gradient. Above, the population has a bell-shaped distribution along a microsite gradient, but only during periods of most favorable weather will the population occupy the full range of microsites it is potentially able to occupy as indicated by the curve. Below, cumulative numbers from left to right are plotted against the same microsite gradient, forming a sigmoid curve. As weather, as a temporal variable, becomes increasingly unfavorable, the population numbers follow the lower curve down and to the left. If weather fluctuates in a range that is favorable relative to the mean adaptation of the population, population numbers fluctuate in range (a). If weather fluctuates above and below the mean adaptation of the population, population numbers fluctuate more widely in range (b). If weather fluctuates in a range that is predominantly unfavorable relative to the mean adaptation of the population, population numbers fluctuate in range (c); but the population is buffered by the survival of the few individuals in most favorable microsites on the left tail of the curve. (D. Goodman & R. H. Whittaker, unpublished.)

are more than the tail of the population; they are part of its buffering, its protection against extinction. Daniel Goodman and I have worked on a population model based on this concept and found the results suggestive. Let us deal briefly with only one of the variables, the relationship along the microsite gradient of the mean position of the population in a favorable period, and the mean position of the effect on it of the fluctuating temporal variable. Modeling the effect of the temporal variable on the population then produces such traces as the three in Figure 2. They simulate real population behaviors, and some of you familiar with population data might choose to name these species *Antiparus*, *Antilepus* and *Antibupalus*. There are evident limitations on this model that I cannot discuss here. It is of course a simplification that projects onto one axis in an idealized form the variety of different interactions of microenvironmental and temporal variables that affect real populations. I suggest, however, some observations:

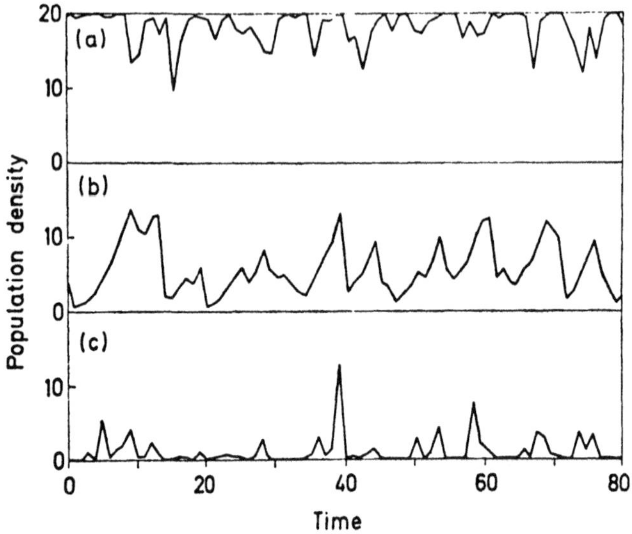

Figure 2. Manners of fluctuation in hypothetical populations modeled on the assumptions of Figure 1. Populations fluctuating in ranges (a), (b) and (c) of Figure 1 are illustrated. (D. Goodman & R. H. Whittaker, unpublished.)

1. What counts in evolutionary time is not whether a species population fluctuates more or less, or whether the species has well-defined density-dependent limits on its increase. What counts is survival (Slobodkin, 1968). What counts for survival in turn is the limitation on the downward fluctuation of the population—buffering mechanisms that reduce the loss from the population in adverse periods and thus prevent extinction. As Smith (1972; cf. Birch, 1971, and Orians, 1975) has observed, microenvironment heterogeneity is prominent among these buffering effects that are fundamental to whatever long-term stability and resilience a community possesses. One means of buffering is the survival of individuals most favored in microsite, or genetic characteristics, or both.

2. To our concept of the community—of species populations dispersed in a microsite hyperspace—we now add fluctuation of environment. It is likely that an unmodified mosaic balance that in theory could permit coexistence of species without niche difference in a stable environment, would not permit that coexistence in a fluctuating environment. As temporal environment changes, however, the populations contract and expand, variously shift back and forth in microsite position, or go into and out of protected life-cycle stages. Thus the steady-state pattern of relationships among populations in the community can persist from decade to decade, while changing numerically from season to season and year to year. Thus the community can persist as a regularly or irregularly pulsating flow system. Thus the community is, in a sense, buffered by the varied buffering mechanisms of its species.

3. We note the diversity of manners of population function. The recognition of r and K selection (MacArthur & Wilson, 1967) seems not enough; we may distinguish along a continuum: (a) selection emphasizing adaptation to other individuals of the same and other species in a relatively favorable and fully occupied environment, saturation- or interaction-selection, (b) selection emphasizing high reproductive rate and dispersal in unstable environments that are intermittently favorable and unfavorable, exploitation-selection, and (c) selection emphasizing survival of periods of hardship in a predominantly unfavorable and restrictive environment that only at times becomes favorable enough to permit population increase, adversity-selection. Figure 2 represents typical population behaviors of species subject to these three types of selection. Saturation-selection may correspond to K-selection and exploitation-selection to r-selection, but we think there is advantage in distinguishing three selective directions rather than two. The terms K- and r-selection may be somewhat misleading, except as mere labels for selection for interaction or exploitation, for it is not simply true that selection maximizes either K or r. Clearly, selective types can be distinguished only by relative emphases, for all species are subject to selection for a sufficient r (and all or most are in some times and places subject to some degree of saturation- and adversity- selection).

4. The species in a community differ in manner of population function, kind of buffering, and degree of fluctuation. Antimaple maintains its population by copious production of seedlings of which a small fraction survive, mainly during infrequent favorable periods; antibeech maintains its population by a sparing, but more steady production of seedlings and suckers. Two parasitoids can coexist, one predominant at the higher and the other at the lower population levels of the host (Utida, 1958). The differences by which species survive in the same community include differences in response to fluctuations of environment and of other species. To the microsite hyperspace we can add axes for time-relations and interactions with other species. The hyperspace with axes for all intensive variables that relate species of a community to one another is the niche hyperspace. By the species' position in the hyperspace, hence its relation to other species in the community as a system of species that live together and may interact, we describe its niche (Hutchinson, 1958; Whittaker, Levin & Root, 1973).

Community differences

Let me now expand this view in one more dimension—the differences in manners of maintenance of dominant species in different communities. Some forests have the kind of continuous turnover of balanced dominant populations I have proposed for antibeech and antimaple. Boreal forests or taiga appear more commonly to reproduce irregularly, every century or two, through successions following fire or blow-down (Jones, 1945). The hard chaparral or California sclerophyll shrubland has a fire cycle; at intervals that may be 20 to 30 years fire burns off the above-ground shoots of the shrubs. Following the fire annual plants grow and bloom profusely for a few years until the shrubs grow back from surviving roots and suppress the herbs (Muller et al., 1968). In grasslands of semiarid climates wide fluctuations of populations are observed, with dominant species falling from major to minor fractions of coverage, or disappearing from some stands and surviving only in most favorable topographic positions (Weaver & Albertson, 1956). The species with the latter behavior is buffered, according to our model, by population displacement along an extensive or habitat, rather than an intensive or microenvironmental, gradient. Many deserts have two major groups of plant species with different adaptations to environmental fluctuation—perennial shrubs that mostly survive the fluctuation with long life and slow growth, and short-lived ephemerals that are buffered against unfavorable environment by their seeds and germinate and grow rapidly to maturity in some years when the rainy season gives them a good chance of reproduction (Whittaker, 1972). In relation to the model the ephemerals of both chaparral and desert are adversity-selected. These differences among communities may be familiar, but they suggest some points:

1. Communities differ widely in the kind and extent of their population fluctuations, and consequently in their degrees of stability.

2. These differences are to some extent interpretable as adaptations to different kinds of environmental fluctuation. This suggestion of community-level adaptation we should qualify in two ways. First, it is difference in population behavior of dominant species that we are actually observing. Second, these kinds of adaptation are flexible; it is quite possible to have annual grassland and chaparral side by side in the same climate, with very different population responses to the same climatic fluctuations. We might better say that the different manners of population function are joint products of kinds of environments and kinds of dominant species that have evolved and form communities in those environments.

3. Though we are describing behavior of dominant species, that behavior must have significant effects on other organisms of the community. The implications of different plant community function for animal populations have not much influenced thought on animal population dynamics, apart from some studies of the tundra (Schultz, 1964).

4. We note again the diversity of behavior of different species populations in a given community. The contrasts of shrub and herb behavior in the chaparral and desert are only striking cases of the fact that the community response to fluctuation is a complex pattern of different behaviors by different species. Communities differ in the ways climax and succession relate to one another (Whittaker, 1974) and in the extent to which localized disturbances and small-scale successions, within a larger and relatively stable community matrix, are part of their normal community function and a basis of their diversity of species (Levin & Paine, 1974, 1975).

Stability

Let me come now to the question of stability in relation to diversity and maturity that is our theme. Orians (1975) has ably distinguished types of stability, emphasizing functional concepts, but let me go further, emphasizing the descriptive aspects. Constancy is unitary, but departures from constancy are various. For each variable that concerns us we can choose to apply stability-instability to: (a) relative amplitude of a regular fluctuation, (b) relative irregularity of fluctuation and (c) presence of zero values in the fluctuation. Furthermore, given a pattern of fluctuation with any combination of these, we can choose to regard as a stability (d) the duration of this pattern in evolutionary time. These four qualities of fluctuation and duration can be applied both to the community and to environment. This implies eight aspects of stability, to which we must add the ability of the community or the several-species system to recover from perturbation—this being the preferred definition of May (1971, 1973), Holling (1973), Orians (1975) and others. This is nine, and I am two up on Orians without factoring the functional aspects as he has done. I would also suggest that the four qualities of fluctuation and their duration can be applied on the three levels of environment, individual populations and communities; but these twelve combinations plus recovery from perturbation would give me an unwelcome total number of kinds of stability. Let me focus, for my purposes, on three of these. Population survival of environmental fluctuation depends in large part on what I have called buffering. Communities are in a sense collections of species that have not yet become extinct because their buffering or tolerance of fluctuation has made possible survival of the environmental changes (and biological pressures) they have so far encountered. Relative amplitude of population fluctuation seems a less fundamental and more variable thing. We expect species to reach different kinds of adaptive trade-offs implying different intrinsic rates of increase and different degrees of population fluctuation. We should expect wide differences among species in amplitude of fluctuation within communities, and among dominants of different communities. The third aspect of stability is the relative constancy of the community's over-all metabolism or productivity from year to year. This we may think determined primarily by resource flow—of water and nutrients especially—as governed by availability, temperature as an external factor, and characteristics of the community itself. I would suggest these further observations:

1. These three kinds of stability have no necessary relation to one another. When we talk about stability in ecosystems we are talking about congeries of phenomena. This is not to say these different aspects of stability are wholly unrelated—we have suggested, for example, that resource flows are, on different levels, part of the basis of relative stability of both community function and some individual species populations.

2. Species stability, whether defined by amplitude or survival, appears to be

dependent on the population mechanisms and buffering of particular species, or the interactions and dependences of small groups of species. One may think of the whole community as comprising loosely interlinked component communities, each including a plant species, the herbivores and symbionts adapted to the chemistry of that plant species, and other species interacting with these herbivores and symbionts (Root, 1973). Although the component communities are variously linked by shared species, each may be to a degree self-regulating. Increased diversity of the community as a whole need not imply increased stability of individual species and groups of interacting species.

3. Environmental stability has ambivalent meaning in relation to species diversity. We expect a stable, favorable, equable environment to permit the evolution of high species diversity, in which many saturation-adapted species are accommodated to one another by narrow niche specialization and some of these have relatively stable populations. We observe on the other hand that a desert environment that is unstable in senses (a) to (c)—one with precipitation with wide amplitude and irregular fluctuation including zero values—but has been of long duration in sense (d) can support communities of high species diversity (Whittaker, 1972). It appears that niche differentiation in the desert involves difference in adaptation to environmental fluctuation, and that environmental instability has been part of the basis of evolution of the present high plant species diversity of this community.

4. It seems difficult to apply to whole communities the stability criterion of re-recovery from perturbation, as May (1975) observes, or the further concepts discussed by Orians (1975). High diversity may imply complexity of interactions among species and a strong role of biotic influences on the microsite mosaic. It is reasonable to suppose that these complex and sensitive relationships, once raveled by disturbance, would be reconstituted less easily than the simpler relationships of less diverse communities. Complex communities may be particularly vulnerable to disturbances such as they have not experienced in their evolution; they may be dynamically fragile (May, 1973, 1975). It seems also, however, that less diverse communities of extreme environments—desert and tundra—are easily disturbed and slow to recover because their low productivities limit their capacities for regrowth. It does not contradict May's conclusion on the effect of complexity on stability, if other community characteristics also affect stability.

Conclusion

My second point in the last set is a virtual paraphrase of May (1971, 1973), and I am surely in agreement with Orians (1975) on the need for effective analysis of stability concepts. Having built this bridge of concepts and inferences from populations to communities, I find myself in welcome agreement with the major conclusions, from different approaches, of the other members of this symposium and with the rejection of a simple diversity-stability relationship by Maynard

Smith (1974) and Goodman (1975). My own interest, however, is not so much with the stability question as its underpinning in community theory. We want to understand the structural-functional design of the community, on the basis of which we can approach questions of stability and diversity. This seems to be a quite distinctive, and loosely organized, design (Whittaker, 1957, 1969; Whittaker & Woodwell, 1972). The view of the community I have developed here can be summarized in a few points: (a) the role of microsite occupation and modified mosaic balances in governing plant populations, (b) the further contributions of resource flows and differentiation in resource use, and of predation and other species interactions, to the determination of these balances, (c) the significance of buffering mechanisms for species survival as the most essential aspect of community stability, (d) the importance of microenvironmental heterogeneity for population behavior and buffering, (e) the evolution of species in a community toward different manners of population function and degrees of relative population stability, so that the community is a mixture of differently fluctuating populations, and (f) the evolution of communities toward different manners of population function of dominant species, different ways of relating to environmental fluctuation, and different roles of disturbance and succession. I should like to regard these points not as members of a list, but as interrelated elements of a design. It is this conception I offer as a suggestion for research. I have called my example species antibeech and antimaple to emphasize that the design is made largely of hypotheses and speculation, if not of antimatter. I hope when this suggestion encounters reality, when research brings antibeech and antimaple in contact with real beech and maple, the result will not be an explosive dematerialization of the bridge I have sketched.

References

Birch, L. C. 1971. The role of environmental heterogeneity and genetical heterogeneity in determining distribution and abundance. In: Dynamics of Populations: Proceedings of the Advanced Study Institute, Oosterbeek, 1970. P. J. den Boer & G. R. Gradwell eds. pp. 109–128. Pudoc, Wageningen.

Cohen, J. E. 1968. Alternate derivations of a species-abundance relation. Amer. Nat. 102: 165–172.

Connell, J. H. 1971. On the role of natural enemies in preventing competitive exclusion in some marine animals and in rain forest trees. In: Dynamics of Populations: Proceedings of the Advanced Study Institute, Oosterbeek, 1970. P. J. den Boer & G. R. Gradwell eds. pp. 298-312. Pudoc, Wageningen.

Goodman, D. 1975. The theory of diversity-stability relationships in ecology. Quart. Rev. Ecol. (in press).

Harper, J. L. 1967. A Darwinian approach to plant ecology. J. Ecol. 55: 247–270.

Harper, J. L. 1969. The role of predation in vegetational diversity. Brookhaven Symp. Biol. 22: 48–62.

Harper, J. L. and J. White 1971. The dynamics of plant populations. In: Dynamics of Populations: Proceedings of the Advanced Study Institute, Oosterbeek, 1970. P. J. den Boer & G. R. Gradwell eds. pp. 41–63. Pudoc, Wageningen.

Harper, J. L. and J. White. 1974. The demography of plants. Ann. Rev. Ecol. Syst. 5: 419–463.
Holling, C. S. 1973. Resilience and stability of ecological systems. Ann. Rev. Ecol. Syst. 4: 1–23.
Horn, H. S. and R. H. MacArthur. 1972. Competition among fugitive species in a harlequin environment. Ecology 53: 749–752.
Hutchinson, G. E. 1958. Concluding remarks. Cold Spring Harbor Symp. Quant. Biol. 22: 415–427.
Janzen, D. H. 1970. Herbivores and the number of tree species in tropical forests. Amer. Nat. 104: 501–528.
Jones, E. W. 1945. The structure and reproduction of the virgin forests of the North Temperate Zone. New Phytol. 44: 130–148.
Levin, S. A. 1974. Dispersion and population interactions. Amer. Nat. 108: 207–228.
Levin, S. A. and R. T. Paine. 1974. Disturbance, patch formation, and community structure. Proc. Natn. Acad. Sci. U.S.A. 71: 2744–2747.
Levin, S. A. and R. T. Paine. 1975. The role of disturbance in models of community structure. In: Ecosystem Analysis and Prediction: Proceedings of a Conference on Ecosystems, Alta, Utah, 1974. S. A. Levin ed. pp. 56–67. Society for Industrial and Applied Mathematics, Philadelphia.
MacArthur, R. H. and E. O. Wilson. 1967. The Theory of Island Biogeography. Princeton Univ., Monogr. Popu. Biol. 1: 1–203.
May, R. M. 1971. Stability in multispecies community models. Math. Biosci. 12: 59–79.
May, R. M. 1973. Stability and Complexity in Model Ecosystems. Princeton Univ., Mongr. Popu. Biol. 6: 1–235.
May, R. M. 1975. Stability in ecosystems: some comments. In: Unifying Concepts in Ecology. W. H. van Dobben & R. H. Lowe-McConnell eds. Junk, The Hague & Pudoc, Wageningen
Maynard Smith, J. 1974. Models in Ecology. Cambridge University Press. 146 pp.
Muller, C. H., R. B. Hanawalt and J. B. McPherson. 1968. Allelopathic control of herb growth in the fire cycle of California chaparral. Bull. Torrey Bot. Club 95: 225–231.
Orians, G. H. 1975. Diversity, stability and maturity in natural ecosystems. In: Unifying Concepts in Ecology. W. H. van Dobben & R. H. Lowe-McConnell eds. Junk, The Hague & Pudoc, Wageningen.
Root, R. B. 1973. Organization of a plant-arthropod association in simple and diverse habitats: the fauna of collards (*Brassica oleracea*). Ecol. Monogr. 43: 95–124.
Schultz, A. M. 1964. The nutrient-recovery hypothesis for arctic microtine cycles. II. Brit. Ecol. Soc. Symp. 4: 57–68.
Skellam, J. G. 1951. Random dispersal in theoretical populations. Biometrika 38: 196–218.
Slobodkin, L. B. 1968. Toward a predictive theory of evolution. In: Population Biology and Evolution. R. C. Lewontin ed. pp. 187–205. Syracuse Univ.
Smith, F. E. 1972. Spatial heterogeneity, stability, and diversity in ecosystems. In: Growth by Intussusception. E. S. Deevey ed. Trans. Conn. Acad. Arts Sci. 44: 309–335.
Utida, S. 1958. Population fluctuation, an experimental and theoretical approach. Cold Spring Harbor Symp. Quant. Biol. 22: 139–151.
Weaver, J. E. and F. W. Albertson. 1956. Grasslands of the Great Plains: Their Nature and Use. Johnsen, Lincoln, Nebraska. 395 pp.

Whittaker, R. H. 1957. Recent evolution of ecological concepts in relation to the eastern forests of North America. Amer. J. Bot. 44: 197–206.
Whittaker, R. H. 1969. Evolution of diversity in plant communities. Brookhaven Symp. Biol. 22: 178–196.
Whittaker, R. H. 1972. Evolution and measurement of species diversity. Taxon 21: 213–251.
Whittaker, R. H. 1974. Climax concepts and recognition. In: Vegetation Dynamics. R. Knapp ed. Handb. Veget. Sci. 8: 137–154. Junk, The Hague.
Whittaker, R. H., S. A. Levin and R. B. Root. 1973. Niche, habitat, and ecotope. Amer. Nat. 107: 321–338.
Whittaker, R. H. and G. M. Woodwell. 1972. Evolution of natural communities. In Ecosystem Structure and Function. J. A. Wiens ed. Oregon St. Univ., Ann. Biol. Colloq. 31: 137–159.

Author's address:

R. H. Whittaker
Ecology and Systematics
Cornell University
Ithaca, New York 14853
U.S.A.

Discussion

Summarized by H. Klomp

Participants: the authors G. H. Orians (U.S.A.), R. Margalef (Spain), R. M. May (U.S.A.), R. H. Whittaker (U.S.A.), the chairman V. Westhoff (The Netherlands), together with E. Halfon (U.S.A.), G. P. Harris (Canada), S. Nilsson (Sweden), B. C. Patten (U.S.A.), T. S. R. Schütt (Sweden).

Elasticity, defined as rate of change of recovery of a factor after perturbation, has not a wide applicability because of the difficulty of measuring such rate of change. Moreover, it does not indicate any oscillations that occur while returning to the original state. The system may overshoot this original state, and the velocity of recovery may change with time. Thus, is this measure a number or a series of numbers? *(Halfon to Orians)*.

I do not believe that measurement of the rate of change is the critical problem, because there are many mathematical techniques for measuring rates of change. Rather, our problem concerns what features are really the relevant ones to measure. As an example we can consider the results of one of the most important recent perturbation experiments, that of Simberloff and Wilson in the Florida keys. When they defaunated islands, they found that the number of species returned quite rapidly to the previous values. Therefore, there was high elasticity in the species richness component of the communities. Nevertheless, the actual species compositions of the new communities were often very different from the original ones, and a taxonomically-oriented ecologist would have measured a much lower rate of elasticity.

These same data were subsequently reanalysed by Heatwole and Levins who showed that the new communities were actually very similar in their trophic structures. Thus, trophic elasticity was high.

All three of these measures are useful in understanding different kinds of problems and it is pointless to debate whether one is better than another. Any measure of elasticity has meaning only in the context of some hypothesis. It is the formulation and elaboration of these hypotheses that constitutes the real problem. Once we know what we want to measure, we will probably be able to measure it. *(Orians)*.

I agree that the concept of stability is a complex one. Non-linear systems can have several kinds of stability, e.g. constancy, limit cycles, rigidity (inertia), etc. Some of the system variables may have stability, while others have not. There may be several equilibrium points or trajectories, and the conditions for stability may change with the forcing functions (input). Hence, many possibilities exist for a given system. Then, the proper meaning of a stable system would be that all its variables are stable in all equilibrium points or trajectories, and for all proper sets of inputs. I believe that more diverse systems utilize the environment more efficiently than less diverse ones, and that the former are more fragile when environmental conditions change. Therefore, the question should not be: 'Is a more diverse system more stable than a less diverse one?' but instead: 'How stable is the diversity in a given system?' (*Schütt*).

It is doubtful whether there is a simple relationship between the diversity of a system and the efficiency of resource utilization. Among plants, for example, where all species use the same resources in the same manner, monocultures appear to use resources as efficiently as mixed species associations. Agriculture is based on this fact. Among animals, however, where there is more variation in resources used, and when, where, and how they are used, mixtures of species usually use resources more completely. (*Orians*).

Ecology does not need 7 or 8 stability concepts. One can do, in its dual aspects, with resistance (constancy) and resilience (return after perturbation). Redundancies in Orians' scheme are as follows: cyclic and trajectory stability are the same, particularly for periodic trajectories; Orians has graphed the former in dimension-1 × dimension-2, the latter in time × dimension. Persistence relates to rate of parameter evolution; if the rate is slow or zero the system persists, if it is fast it changes rapidly, but never in an unstable way. Inertial stability may be viewed as a force (cause), which produces constancy or persistance as effects. Amplitude stability simply measures the magnitude of the state space, given that multiple attractive basins are not possible (as asserted in my paper for the Congress on the zero-state). The real value of Orians' stability concepts lies in their great potential as measures of relative stability, given that life and ecosystem are globally stable. (*Patten*).

Considering stability of complex ecological systems, it will be fruitful to recognize the homologous nature of genetic and physiological systems. To what extent do your models inform us about the genetic, evolutionary, and physiological properties of individuals and populations? (*Harris*).

Yes, it would be fruitful. The task of extending my type of studies (which are evolutionarily static) to include the effects of evolutionary adaptation to change and disturbance, is a relevant but difficult one. (*May*).

I agree that we finally need a unifying theory to explain the pattern of global diversity. However, for the time being, it may be a more proper strategy to try to develop hypotheses about conditions affecting diversity within limited groups of organisms restricted to a limited area. (*Nilsson*).

I agree. I think my paper says this, and gives some references. MacArthur's book 'Geographical Ecology' (Harper & Row, New York 1973) gives many others. (*May*).

Session 4

Diversity, stability and maturity in ecosystems influenced by human activities

Chairman: G. E. Likens

Diversity, stability and maturity in ecosystems influenced by human activities

Jürgen Jacobs

Introduction

In discussions of the impact of man on ecosystems it is usually argued that man interferes with mature equilibrated situations, thereby reducing diversity and destroying inherent mechanisms of stability. In this argument it is tacitly assumed that (1) the majority of ecosystems are undisturbed and more or less mature prior to human intervention, (2) mature systems are more complex and stable than immature systems, and (3) man acts as an external, unnatural force on natural ecosystems. I want to try to show that these implications are not necessarily correct or at least very one-sided. I shall present some selected evidence that man affects diversity, stability, maturity and organization in all possible directions and quantities, and that man, though unique, is to be regarded as an integrated, natural component of ecosystems, if his role is to be fully understood.

Discussions here have made it clear that various definitions are used for the loaded words diversity, stability, maturity and organization, and that the biological meaning of some of them may be multifactorial, obscure or even absent, even though they may yield predictive correlations in field data analysis. Therefore, I shall briefly define the meaning of such terms for the duration of my talk and state some of the frequently assumed correlations.

Definitions

Diversity will be used in a *descriptive, static way*, as defined by the Shannon-Wiener formula or any equivalent. It contains two components, namely *richness or number* of groups, and the *evenness* of distributions of the relative abundances of the individuals within each group (Pielou, 1966). Both components may refer to species, or other groups of organisms, or to any other quantifiable units in the ecosystem. Diversity is never a measure of whole ecosystems but only of certain components or aspects thereof. It is assumed that the term is a hazy reflection of multiple, dynamic functions in the system.

Stability, unless otherwise specified, is also used in a descriptive sense, as persistence of a given state or range of states in time (cf. Margalef, 1969). Nothing particular is implied *a priori* as to the causes contributing to stability, but *complex structures* at all levels of organization which reduce the turnover of matter or energy per biomass, and *equilibrating negative feedbacks* are accepted as major contributors.

The term *maturity* shall apply to the successional stage of an undisturbed ecosystem gradually approaching a more or less stable climactic stage. Maturity does not have to increase indefinitely but may be followed by decay of the ecosystem. There may also be a sequence of different ecosystems such as in the change from lake to forest. The term maturity is as vague as are the rules of succession.

With *organization* I mean amount of order, that is, deviation from a random situation. I do not know how to measure the amount of organization in a natural ecosystem in a comprehensive way. The problem of how the multiplicity of all organizational qualities of an ecosystem is translated into quantitative units, is still entirely unsolved. So I shall often have to stick to the intuitive use of the words organization and structure.

Correlations between diversity, stability and maturity

There are numerous correlations between the mentioned parameters but few valid rules about the correlations. Some of these correlations I should like to state as a framework for this talk.

First we have to consider the relations in connection with the *age of the ecosystem*. Diversity always increases in initial stages of succession. This has, at least in part, tautological reasons: it can be shown that both components of diversity, richness and evenness, may increase in a succession which starts from nothing, for mere statistical reasons (Fig. 1). In later stages of succession the the fate of diversity is not predictable at all. Richness and evenness may both increase, or only one of them. There may also be a strong decline of both components (Fig. 2, 3 and 4). Also the sequence of stability and organization in developing systems is all but certain. To be sure, the climax is by definition more stable than are more immature stages if one uses Margalef's measure of stability (Margalef, 1969) but I am unaware of sufficient data on the actual course of stability or organization during succession to derive even a rule of thumb.

Apart from successional correlations, the following major factors influence, or are correlated with, diversity, stability and organization:

1. Spatial environmental heterogeneity usually increases species diversity; the effect on organization and stability is ambiguous.

2. Temporal environmental heterogeneity may increase or decrease diversity and stability, depending on severity, total duration etc.

3. Environmental stress, that is the harshness of environmental conditions, is very often negatively correlated with diversity.

4. Competition on a short-term basis may decrease diversity, due to competitive exclusion. However, with sufficient time for evolution, competition may increase diversity (Wilson, 1969).

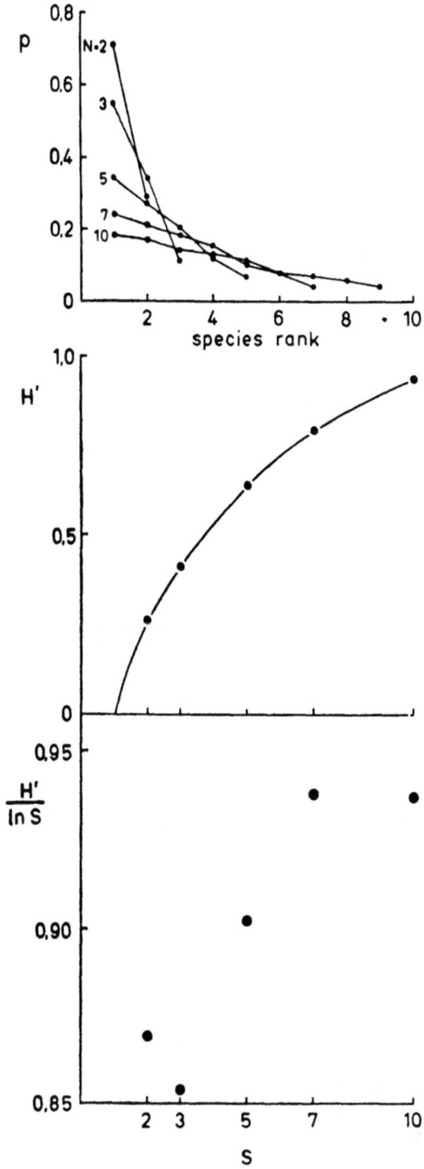

Figure 1. Influence of species number S on relative abundance p, diversity $H' = -\sum p \cdot \ln p$, and evenness $H'/\ln S$. 'Population sizes' were taken from random numbers in the range from 0 to 99. Each value represents the average of 20 replicates.

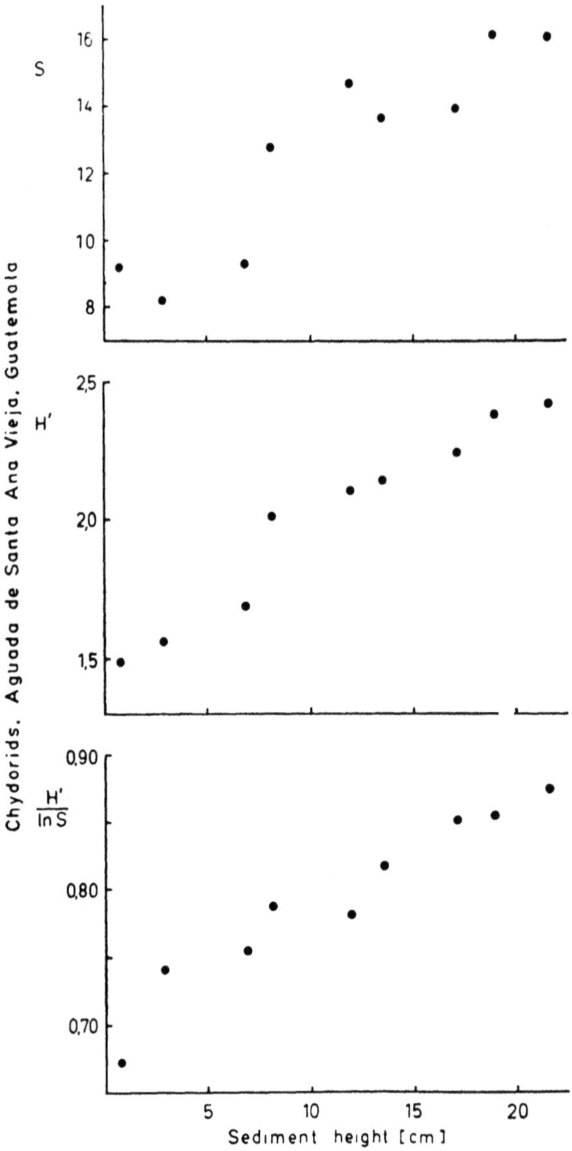

Figure 2. Succession of chydorid cladoceran species number S, diversity $H' = -\sum p \cdot \ln p$, and evenness $H'/\ln S$ in the Aguada de Santa Aña Vieja, Guatemala, as revealed from sediment analysis. Total time of succession about 200–250 years. All three indices increase with increasing age of succession. (Data from Goulden, 1969.)

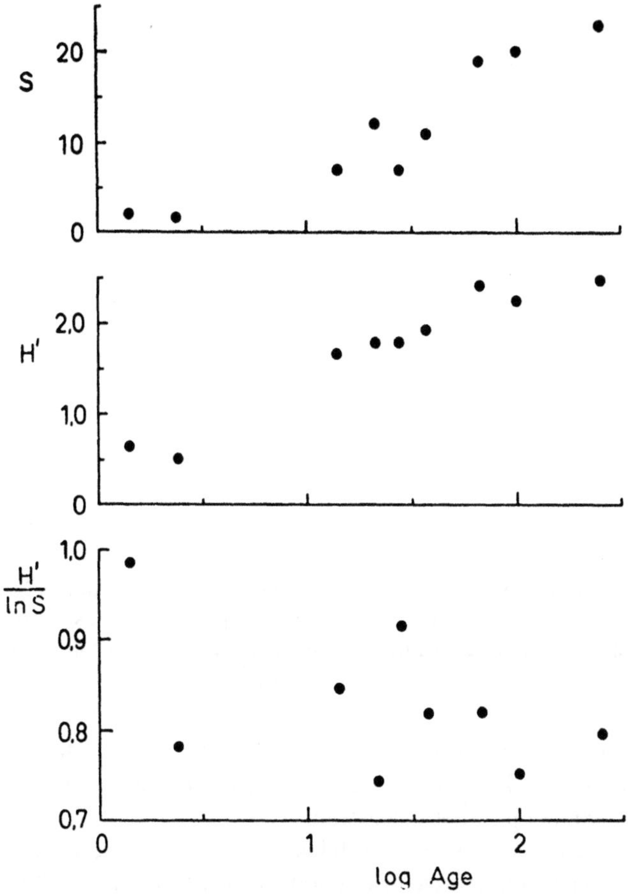

Figure 3. Succession of bird species number S (breeding pairs per 40.5 ha), diversity H' and evenness $H'/\ln S$, in abandoned agricultural fields in the southeastern United States. S and H' increase, $H'/\ln S$ decreases with increasing age of succession. (Data from Johnston and Odum, 1956.)

5. Enemies, similar to competition, may act both ways depending on their strength of action, evolutionary time, and their effect on competition among the prey species.

6. Changes in the amount of energy and nutritional resources are very important, but again, the effects are variable as an example will demonstrate (Fig. 7).

7. The exchange of organisms or system components with other ecosystems may evoke just about every conceivable effect.

Ecological characterization of man

The question is now which role man plays in this uncertain web of factors. As already mentioned, man's activities influence just about every aspect, and in

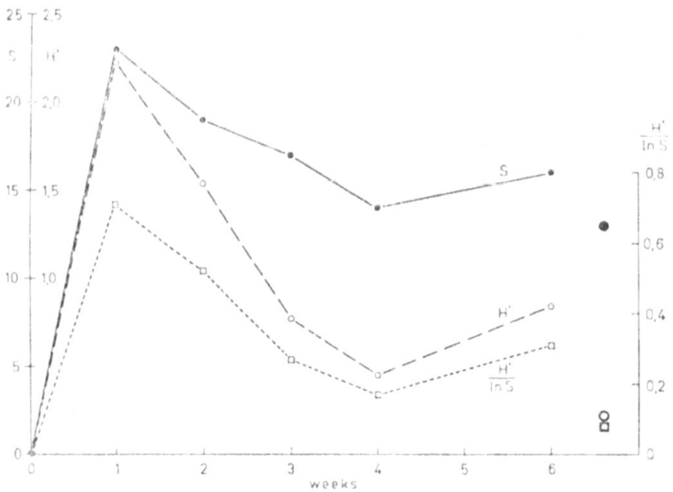

Figure 4. Succession of species number S, diversity H' and evenness $H'/\ln S$, of benthic algae in the Ticino river, Italy, measured on exposed microscope slides. The symbols on the right indicate values from natural benthic substrate. All three indices decrease after the initial increase. (Data from Cattaneo and Ghittori, 1975.)

every imaginable direction. It may, of course, be questioned whether I see the rules or the exceptions. In the latter case, I hope that the evidence for true relations, if known, will come up in the discussion.

The often dominant role of man has to do with his *ecological characterization.* Therefore I should now like to summarize briefly some properties of man which I think are relevant in this context.

1. There seems to be no doubt that our species has the highest biomass of any animal species, some 100 million tons of dry weight, or the equivalent of 6×10^{14} kcal. Man has also a long generation time compared to other animals. Thus the actual population biomass turnover is very slow, in other words, there is an inherent component of stability in our life cycle.

2. Our population, viewed globally, is growing at a superexponential rate of increase, at present somewhat more than 2% per year.

3. The structural organization of human populations is certainly the highest in the animal kingdom. This applies not only to the social structure *within* populations, from families to nations, but also to the unique *inter*-populational relations: most human populations differ from each other, due to divergent evolution of traditions, even more than animal populations belonging to different species. Quite often, the relations resemble enemy-prey or host-parasite conditions. On the other hand, the exchange and interdependence between human populations is, at least now, much greater than is typical for other species. This

concerns the exchange of people as well as that of energy, matter, knowledge, and traditions. This interchange degrades our populations in many respects to true subpopulations.

4. One reason for this presently high level of human organization and diversity is the extraordinary use of *energy* by man. Man does not follow the general mammalian relation between energy output and bodyweight. Instead of some 2–3000 kcal/day · person which would be typical for an animal of comparable size, we have, on the global average, an energy expenditure of about 40,000 kcal/day · person, that is 15–20 times as much. Man is, in this respect, equivalent to an animal having a weight of about 5–7 tons. On the other hand, if we look at the energy consumption per biomass, man behaves like a mammal of about $1\frac{1}{2}$ gram bodyweight. If the total utilization of energy is considered as gross production, then the net production of the global human population is only about 0.2 per mille of its gross production. The latter (about 1.4×10^{14} kcal/day) presents about 1% of global net primary production, or about the same absolute amount as the sun's energy available to the first carnivore level. Our energy expenditure per person at present increases at about 2% per year. Some 90–95% of this energy is not derived from active ecosystems but stems from the energy surplus of earlier ecosystems and, to a small extent, from nuclear energy.

Major effects of man in ecosystems

I should now like to give a very rough survey of the major effects of man in ecosystems. In the face of the multitudes of possibilities, such a survey has to be highly selective and subjective, and is bound to be superficial, not only in the selection of examples but also with regard to the order of arranging the vast spectrum. I shall make four sections: (1) *transient perturbations*, (2) *chronic shifts* of environmental conditions, (3) *energy and nutrient relations* (actually a subtitle to 1 and 2, but given a section of its own because of its importance) and (4) *manipulation of species*.

Transient perturbations

Here we have to consider events such as accidental oil spills, the chemical poisoning of rivers, the non-accidental chemical forest defoliation in warfare, the burning of savannas by african tribes and farmers, or a temporary supply of nutrients. One thing seems to be common to such disturbances: they reset the clock of succession, usually to an earlier, less mature stage, depending on the severity of the perturbations. Considering how ambiguously diversity and succession are related, it is no surprise to find that the diversity of the effects of perturbations is also considerable.

The first example, taken from Odum's text book, shows data from Barrett (1969) on the effects of a sudden insecticide application on the arthropod fauna of a millet field (Fig. 5). Obviously richness and evenness do not go in parallel. In richness, a fast decline is followed by fast recovery, whereas in evenness there is first a slow increase and then a slow return. Diversity, the resultant of both components, shows a corresponding zigzag, first a decline due to richness, then an overshoot due to evenness.

The next example relates stability and succession (Table 1): Hurd et al. (1971) applied a single fertilization perturbation to each of two stages, 6 and 17 years old, in the succession of abandoned hayfields. The amount of the reaction of the ecosystem to this perturbation was taken as a measure of instability. It turned out that the effects of perturbation and hence, stability, were different not only in

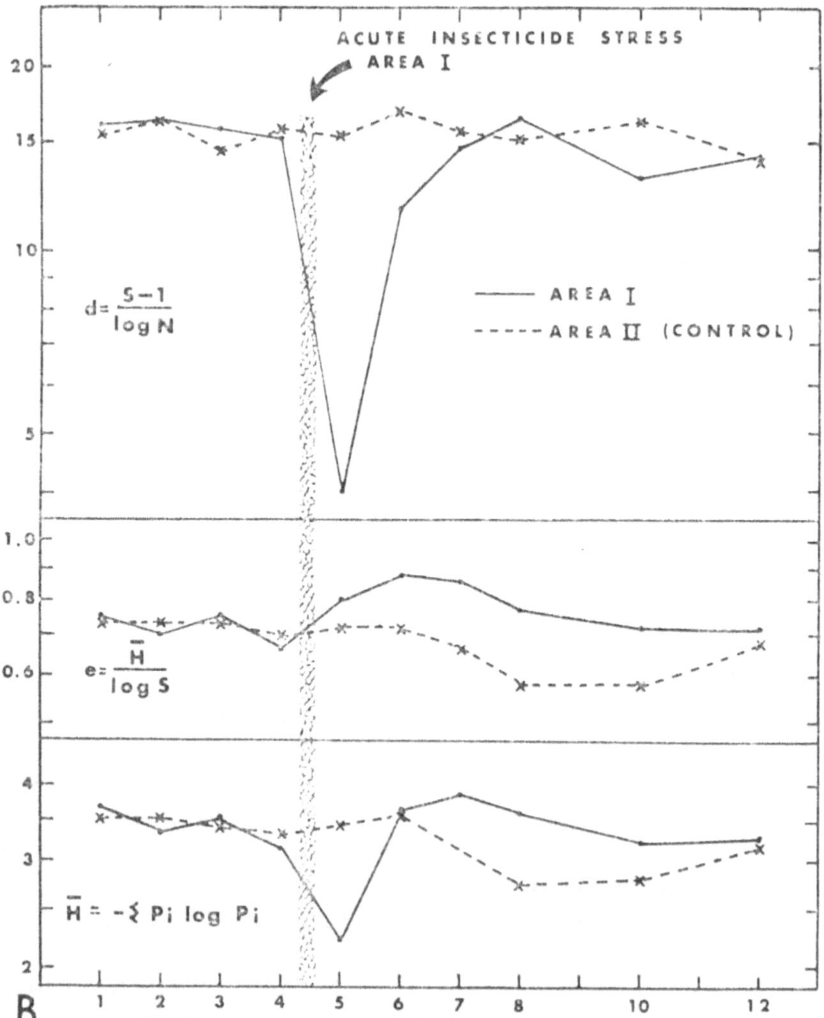

Figure 5. Influence of a temporary experimental insecticide stress on the arthropod fauna of a millet field. For explanation see text. (From Odum 1971, data of Barrett, 1969.)

Table 1. Effects of fertilizer perturbation on plant, herbivore and carnivore production and diversity, in a young (6 year) and an old (17 year) successional stage of abandoned hayfields (data from Hurd et al., 1971).

	Net production			Diversity		
	young	old	Δ	young	old	Δ
Plants	+66%	+22%	+44%	—	—	—
Herbivores	+22%	+207%	−185%	+43%	+20%	+23%
Carnivores	−17%	+72%	−89%	+6%	+28%	−22%

different successional stages but even in the same stage, when different parameters were compared. Concerning *primary production*, the young stage (6 years old) responded much more strongly than the old stage indicating, as expected, low buffering capacity in early succession. At the *herbivore* and the *carnivore* levels, however, the effects were opposite: the younger stage responded much less, thus was more stable. Looking at *species diversity*, the result was equivocal. In all cases, diversity increased, but there was no clear correlation with succession. In the young stage, the stability of *herbivore* diversity was *smaller* than in the old stage, but for carnivores, the situation was just the reverse.

In these two examples, the perturbations were eventually overcome, the ecosystems returning to the original state. However, the re-setting of succession which is caused by a perturbation, may also switch the system into an entirely new direction once certain thresholds are passed. A devastating example is the permanent deterioration of forests, in Italy due to tree cutting and in Vietnam due to chemical defoliation, both leading to irrevocable soil erosion after the perturbation.

Summarizing I conclude that there is no general statement possible as to the effects that are caused by transient disturbances.

Chronic changes

The second type of human activity deals with *long-term*, *chronic shifts* of the environment which means that permanent new levels of conditions are set, or that continuous gradual changes, including accumulations, take place.

Here we have to include a vast array of causes. One could mention *simple* cases such as the continuous or increasing application of pesticides, the increasing emission of lead by cars, or of other waste products and poisons by industry, but also 'positive' changes such as irrigation. More typical for human influences are *complex patterns* of influence in space and time. Such influences comprise, for instance, the regulation of river beds, the multiple treatment of soil involved in the conversion from virgin to agricultural land, or the creation of entirely new structural units, like polders, water reservoirs, and cities.

My aim here is again to point out that there is just no general statement possible as to how such chronic, man-made developments influence diversity, stability etc. A first example (Fig. 6) shows the effects of chronic pollution by a

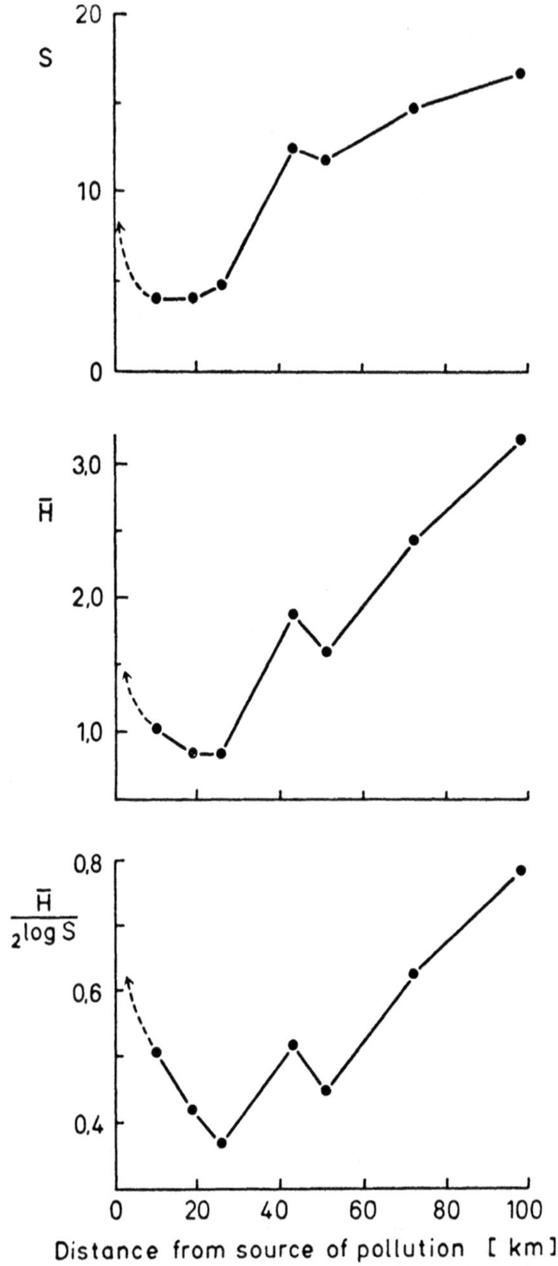

Figure 6. Effects of chronic pollution (mixed home and industrial waste) from a small city, on species number S, diversity $\bar{H} = -\sum p \cdot {}_2\log p$ and evenness $\bar{H}/{}_2\log S$, of the benthic macroinvertebrates of a stream. All three indices show a decline toward the source of pollution. (Data from Wilhm 1967.)

small city (mixed home and industrial waste), on the benthic organisms of a stream (Wilhm, 1967). This is a very straightforward case: all three indices, diversity, richness and evenness, show a decline toward the source of pollution. Hence there is a general negative correlation with the amount of pollution and the benthic organisms conform to habitual ecological thinking. In a study by Maurer (1974) on the effects of lead pollution by cars on the arthropod fauna adjacent to roads, the situation was not quite as simple. Maurer could show that species richness was decreased significantly in correlation with the lead contents of soil and arthropods, but evenness was not, which shows that these two components may or may not vary together under the impact of chronic pollution.

Table 2. Dependence of bird biomass and diversity on urbanization (data from Nuorteva 1971.)

	City (Helsinki)	Near rural houses	Uninhabited forest
biomass (kg/km^2)	213	30	22
No. of birds/km^2	1089	371	297
S	21	80	54
H'	1.13	3.40	3.19
$H'/\ln S$	0.37	0.78	0.80

The next example demonstrates the influence of urban development on species production and diversity: Nuorteva (1971) studied the bird fauna in the city of Helsinki, in agricultural areas near rural houses, and in uninhabited forests in Finland (Table 2). The city supported by far the highest biomass and the highest number of birds (and hence, certainly also the highest productivity of birds) but exhibited the lowest number of species and the lowest evenness and hence, diversity. This would be in accord with a high production/low diversity rule. But the man-made rural areas contradict the rule. Here, the number of species and hence diversity were much higher than in the uninhabited forest, and so was biomass (and, presumably, productivity). Altogether, human civilization brought about a very significant increase of diversity in the whole area: there were 37 species in city and rural areas that were not found in the forest.

In conclusion to this second point I would say that the multitude of longterm shifts of ecosystems created by man very often reduces species richness and diversity but probably in as many cases increases it, especially where man creates new diverse structures in a formerly uniform habitat, and maintains these structures in a relatively stable state. Surprisingly little seems to be known quantitatively about actual stability, for instance, in cities, irrigation areas, in systems of river regulation and, for that matter, in agricultural areas.

Energy and nutrient relations

Let us now consider man's role in the energy and nutrient relations of ecosystems characterized by *exchange processes* which may be conveniently subdivided into import and export.

Import can be of various kinds. The most obvious type is, of course, inorganic or organic fertilization and pollution, mainly by phosphates, nitrates, trace elements, and energy rich organic compounds. But the heating of rivers and shore waters of oceans by nuclear power plants also constitutes a continuous energy injection into the environment with drastic consequences. Still another, more indirect aspect of energy-supply is organized manipulation: the high energy consumption of the human population has as one of its primary correlates the tremendous organization of technology and power of manipulation. The complex but orderly management, for instance, of agricultural areas is really an emanation of this human energy utilization.

As an example I wish to discuss one well-known case of nutrient import: the pollution history of Lake Washington in the U.S. which has been not only studied but also drastically changed by the efforts of Edmondson (1972 and earlier) and his coworkers (Fig. 7). I shall stick to primary producers.

Let us first look at diatoms which were studied by Stockner and Benson (1967). Araphidinate diatoms are characteristic indicators of eutrophic lakes, whereas the Centrales are typical for oligotrophy. Up to about 1840 there was a low percentage of araphidinate diatoms and a correspondingly high percentage of Centrales. This is also shown by the relative abundance of two representative species, *Fragilaria crotonensis* (Araphidineae) and *Melosira italica* (Centrales). This is then the typical situation for an oligotrophic lake. From 1840 to about 1920, the time of the gradual development of the City of Seattle on the lake, a gradual eutophication evidently took place due to raw sewage released into the lake. This is indicated by an increase in the percentage occurrence of the Araphidineae and by a reversal of *Fragillaria* and *Melosira* abundances. In the following 15–20 years there was a trend to the original oligotrophic state. This corresponded with a gradual termination of all raw sewage effluents between 1930 and 1941. After that, treated sewage was increasingly discharged into the lake again which reversed the situation once more. Actual phosphorus (after 1930) and chlorophyll (after 1950) determinations show clearly this increase of eutrophication. Due to public action the sewage of Seattle was gradually diverted after 1962, which resulted in a strong decrease of phosphates and algae biomass (measured by chlorophyll content). Unfortunately, data on the change in diatom species composition after 1962 have not yet been published.

Now let us look at diversity. It is obvious that there are parallel developments but there are also some clearly unexpected results. During the development of Seattle, eutrophication and hence probably also productivity and biomass, increased. If there was an increase in eutrophication, there should have been a decrease of diversity. But diversity stayed about where it was, suggesting that

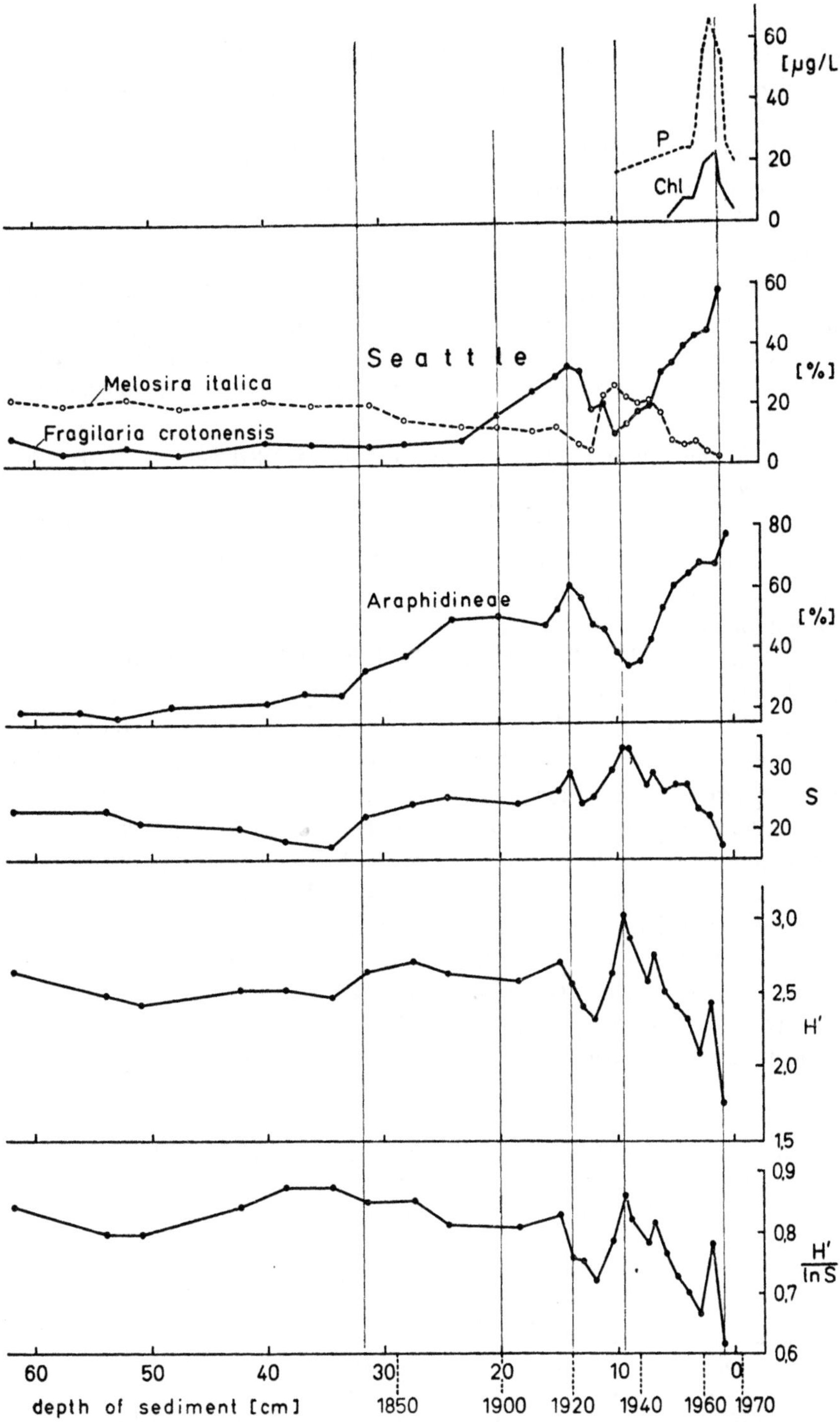

Figure 7. The influence of pollution on the diatom composition of Lake Washington, U.S.A. *P*: total phosphorus concentration in top 10 m. *Chl*: mean chlorophyll concentration in epilimnion. *S*: number of species. $H' = -\sum p \cdot \ln p$. For further explanations, see text. (Data from Stockner & Benson 1967, and Edmondson 1972.)

the system remained fairly stable. A closer examination reveals that this was not true: there was an increase of species numbers but a decrease in evenness. Thus the causes of constant diversity were switched from evenness to richness. The first reversal to oligotrophy which is indicated by a 10–15 year stretch of straight reduction of araphidinate relative abundance, brought first a drop, but then an unexpected increase in diversity. Both components, richness and evenness, participated. After 1940, the situation was straightforward: eutrophication brought a change in composition, a decrease of species richness and of evenness. Species richness in the Fifties and Sixties thus behaved in the opposite way to that in the earlier period of settlement. In conclusion, this example demonstrates clearly that there is no simple rule about the relation between eutrophication and diversity.

Export of energy and nutrients can both be labelled *harvest*. If we take a closer look at harvest, there appears to be a syndrome of correlations: monoculture of the species to be harvested, maximization of production, low maturity (usually used as a synonym for high production), low species diversity, low intrinsic stability but extrinsically manipulated constancy. Usually it is implied that monoculture, low diversity and high yield are causally related. This classical picture is shown in the following example (Table 3). Total arthropod production and diversity were compared in two successive years in a grain field left unharvested (Odum et al., 1971) The first year reflects, except for the actual harvest, the more or less normal state of the monocultural field. The second year shows what happens if monoculture is not maintained by human manipulations: the production strongly decreases, both components of diversity increase.

Actually, however, the relation between yield and diversity does not have to be negative at all. There is no obligatory causal linkage. Rather than to maximize yield, the rational for monoculture may be just the desire to harvest a pure species. Of course, if we specialize fields for a single species as well as for high yield, then we have a linkage. But then the statement that high production means low diversity is a wrong causal statement. In fish ponds where the aim of monoculture is not stringent, it is well established that the total production is greater with more and diverse fish species than with monoculture: in the orient,

Table 3. Total density, species richness, diversity, and evenness of all arthropods in an unharvested grain field and in the successional community of the following year. All differences are significant with $p < 0.01$ (data from Odum et al., 1971).

	Grain field	Successional community	$\Delta(\%)$
Density N/m^2	624	355	−43 %
Richness $(S-1)/\log N$	15.6	30.9	+98 %
Diversity H'	3.26	4.49	+38 %
Evenness $H'/\ln S$	0.68	0.84	+24 %

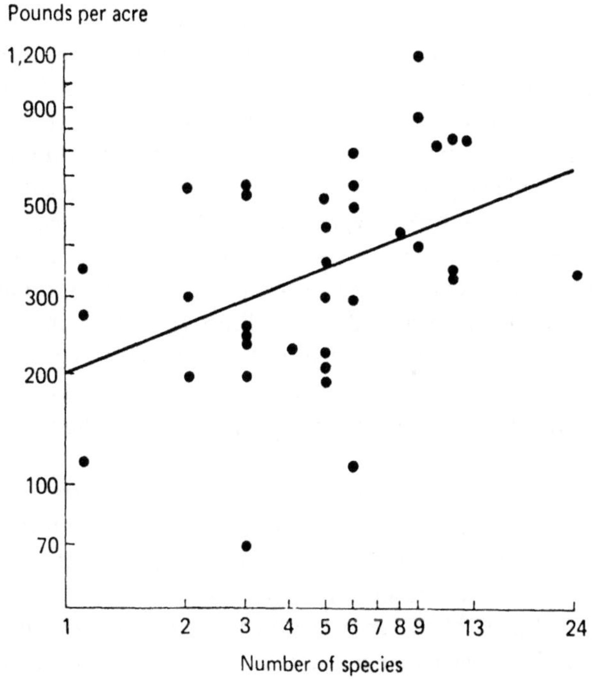

Figure 8. The relation between the number of fish species in U.S. reservoirs and the standing crop (from Watt 1973, data of Carlander, 1955).

fish ponds may be stocked with up to nine carp species to maximize yield (Watt, 1973); Figure 8 shows the positive relation between species richness and standing crop for reservoirs in the U.S.A. (Carlander, 1955).

Human manipulation of species

The final section concerns the *manipulation of species by man.* Analogous to energy relations we may distinguish between species import and export, the latter really meaning extermination. I shall restrict myself to animal import. Accidentally or intentionally, man often introduces species to ecosystems, that is, he facilitates or executes colonization. This applies to beneficial species as well as to pests. It is well known, that new colonists, if they become established, can have profound effects on several trophic levels. Enemy as well as food species and competitors may be drastically affected. The vast literature on the role of 'negative' pest species and the well-known instability they may inflict on a system will not be reviewed here. I want to present two examples that deal in a positive way with richness and stability.

Lake Nakuru is a shallow soda lake in Kenya with extremely simple structure because of the harshness of the alkaline conditions. Prior to 1961, there were essentially only one or two species of algae, one copepod, one rotifer, corixids, notonectids and some 500 000 flamingoes belonging virtually to one species. In

1962 a cichlid fish, *Tilapia grahami*, was introduced in order to check mosquitos. As it turned out, man profited in an unexpected way. There were not really any substantial quantities of mosquito larvae in the lake to feed on. Instead, *Tilapia* established itself as one of the major algivores, reaching a biomass of about $2\frac{1}{2}$ grams of dry weight per square meter. The fish never reached any direct economic value because of their small size. But some 30 species of fish-eating birds, pelicans, anhingas, cormorants, herons, egrets, grebes, terns and fisheagles colonized the area as a consequence of the primary introduction. Lake Nakuru became a new, much more diverse system. It also became a National park and today contributes substantially to tourism and thus indirectly to the economy of Kenya. Inspite of this drastic increase of diversity, there appears to be little influence on *de facto* stability, because it is mainly environmental factors (fluctuations of salt concentration and concomitant parameters) that have the controlling power.

The other example is about the introduction of parasites to control pests. It is well known that introduced species may become pests because the multiple parasite system which helps to control them in the original biotope are not introduced along with the host. One important ingredient of integrated pest control is therefore biological control by secondary introduction of parasites. In our context, one question is of particular interest: is it possible or practical to inject stability by introducing a diversity of several parasites, assuming that diversity of parasitism which is frequently found in nature, increases the buffering capacity of the system? There has been some controversy over this question. Proponents of the one-parasite-thesis argue that competition of several parasites for one host species would lead to competitive exclusion of parasites and hence to an overall reduction of parasite efficiency; multiple parasite systems would only work with multiple host systems (Watt, 1965). There are indeed many cases where one good parasite does the job and where multiple introductions were either without effect or actually may have been harmful. On the other hand, Hassel and Varley (1969) have shown theoretically that multiple parasitism is not only possible but also recommendable if the parasites have a good self-regulatory capacity by mutual interference, and the exploitation efficiencies are not too different. An empirical example is that of the winter moth *Operophtera brumata* which was accidentally introduced from Europe to Canada and became a serious pest species. Nineteen tested Canadian species of parasites were ineffective, but a combination of two european parasites, the tachinid fly *Cyzenis albicans*, and the ichneumonid wasp *Agrypon flaveolatum*, proved to be an almost ideal, highly effective pair (Embree, 1966). There appears to be a division of labour: the wasp is effective at low host densities, it is euryphagous and not a very good exploiter. The fly, on the other hand, is stenophagous and a good exploiter, being effective at high densities of the moth. This example shows that man can stabilize unstable situations by inserting specialized diversity into the system.

Some generalizing statements and conclusions

After this sketchy survey which was intended to show that man influences ecosystems in every imaginable way and that stability, diversity and organization may be changed by man in both directions I shall try to arrive at some tentative generalizing statements.

1. *There is no human effect on ecosystems which could be called typically or exclusively human except by quantity and combination.*

All effects caused by man are, at least in principle, also caused by other species or occur as natural events from abiotic causes. Some examples: primitive tribes such as the Masai, by periodic burning of bush and grassland, are only imitating naturally occurring fires even though they perhaps create a somewhat greater order of events. Specific as well as general poisons are released into the water by algae which may produce massive fish kills. Many antibiotics used by man are borrowed from other organisms; borrowing poisons from others is not uncommon in nature. The addition of inorganic and organic nutrients to estuaries (eutrophication) is the normal consequence of decaying marsh grass. The trend toward natural mono-'cultures' by allelopathic action or competitive shading is well known in plants. Homogeneity of ecosystem structure is produced by territorialism. New ecosystems are created by numerous species, for example oysterbanks or mangrove and papyrus islands. The facilitation of colonization and extermination is a quite natural business of many animal species which act as carriers and enemies respectively. Energy transfer from one system to another is not uncommon: for instance, birds may seek their food in one system, but breed or live in another. What makes human effects human then is not specificity but sheer quantity, variety and combination in mass, space and time.

2. *If the role of man in ecosystems is to be fully understood, he and his social structure has to be treated as a natural though highly specialized component of ecosystems rather than as an unnatural external force.*

This thesis may be illustrated by diversity and stability in agricultural management. Here originally highly diverse ecosystems are often made highly monotonous by man, diversities of plant and animal species are steeply decreased and a tremendous and more or less continuous net increase of production is created. Such a system, according to normal ecological rules, should be unstable, one would tend to call it typically immature. Yet, in reality, high stability is often achieved in such simplified systems by continuous careful and integrated human manipulation. A minor amount of planned diversity and heterogeneity may actually be injected again into such productive systems by means of introduction of useful species, sometimes in a combined fashion, such as pest parasites, by the creation of mixed stands or hedges, by strip harvesting and so on. Thus agricultural systems do not really obey current ideas of correlation between diversity, stability and productivity.

However, the seeming inconsistency appears to be due to the failure to include the properties of human populations. Whenever a system is under the influence of what one may call 'external forces', whatever they are, it would be logical to make these forces plus the causes of such forces a part of the system. In our case, human social organization and all ramifications have to be treated as normal ecosystem components as soon as man enters the system as a causal agent.

At the present time, *intraspecific organization* within ecosystems has not been the object of diversity studies; only interspecific abundance or spatial environmental diversities have been regarded, and even in the latter case in a highly selected fashion. I think this is a mistake, or at least, a very one-sided restriction of the discussion of ecosystem organization. If systems are studied in which primates, including man, play a major role, then social parameters of diversity and stability must be included.

To return to agricultural systems: if human social diversity and stability were measured and included in the assessment, I think an overall positive correlation between diversity, stability and organization would be found.

Similar considerations apply to *energy relations*: man deprives agricultural ecosystems continuously of organic energy plus biomass by harvest to an extent scarcely ever found in manless systems. In return, agricultural areas are supplied with nutrient matter without energy in the form of inorganic fertilizers. Thus agricultural systems must have, almost by definition, a low biomass/productivity ratio. On the other hand, organized human society in distant cities, towns and villages is supported by this harvest. Again, in order to apply ecological rules that are formulated for more or less closed systems, the biomass of man and its energy turnover have to be included in the assessment of agricultural productivity, that is, human population centres have to be regarded as components of a larger system comprising also agricultural subsystems. Actually, the sources of fossil organic energy tapped by man have also to be included because the manipulation of the agricultural subsystem means the injection of highly complex organization into it, and the energy which is necessary to create this organization is derived from fossil fuel. Thus the total system is a very complex, rapidly growing heterotrophic system with a spatial mosaic of auto- and heterotrophic components. In such a system, the biomass/productivity ratio would be quite different from that of the agricultural subsystem alone.

From the foregoing several other interrelated generalizations follow:

3. By the ubiquitous presence of man, *the ecosystems of the biosphere* which were never fully closed but fairly separated from each other as long as human biomass was small, *are now wide open with respect to matter and energy exchange*.

Due to human influences in recent decades, a vast number of systems are now degraded to true subsystems. There is the substantial flow of energy and matter from agricultural areas to urban developments already mentioned, the enrichments of lakes and agricultural lands with fertilizers and energy-rich pollutants, not to speak of the quantitatively minor but qualitatively potent dispersion of pesticides, heavy metals, the introduction of species etc. Above all, there is the tremendous global enrichment of the biosphere by fossil and nuclear energy. The energies derived from these sources, and organization derived from this energy, is being channeled to an increasing number of ecosystems. Thus the various systems are now, by the encompassing activity of sprawling mankind, transformed to a huge system web.

4. Due to man, *the total biosphere is now heterotrophic* in the sense that it lives on organic energy capital of earlier savings of fossil fuel.

This is a reversal from earlier times. At present, about 8–10% of the total energy supplied to the second and third trophic level—this is where man has to be placed—is fossil energy.

5. There is *increasing topographical compartmentalization of the biosphere in subsystems of different characterization and function.*

This concerns the division of labor with regard to the production and consumption of food and energy as well as organization. Urban and industrial subsystems are characterized by heterotrophy, high energy flow and metabolism, high intra-specific diversity and high stability, though perhaps intermediate structural and low species diversity. Agricultural subsystems, on the other hand, are autotrophic and very productive, have a high energy flow, fairly high stability, fairly high organization, but low species diversity, and, at least for some parameters, also low structural diversity.

6. *There is a shift to intraspecific human diversity and stability at the expense of interspecific diversity.*

The pauperization of classical ecosystems with regard to number of species and diversity is counterbalanced by the increasing complexity of the human organization. This present trend of reorganization of the biosphere is based on the increasing percentage of total energy flow in the biosphere being channeled through a single species, man.

7. *Man's manipulative powers make the world's ecosystems increasingly dependent on the structural complexity and integrity of human society.*

Worldwide strategies are increasingly required. Breakdown or inefficiency of human organization will lead to breakdown in the manipulated biosphere. It seems that the functioning of our present organization rests not only on high energy exploitation but on a steady *increase* of that exploitation: economic depressions have profound effects on the state of order in our social systems even if the yearly growth of the gross national product is not arrested but only decelerated. It is obvious that the present trends of increasing energy exploitation cannot go on much longer. But there is no consensus of what will happen to human stability when the present increase of energy flow per person slows down and eventually stops. There are models which lead to certain steady state recommendations, say 400 000 kcal/day · person at a population density of 10 billion people in the year 2050 (Weinberg & Hammond 1971) but such models do not tell how the population can be stopped at 10 billion and how the economic machinery can be induced to accept a steady state system.

8. *There are virtually irreconcilable conflicts of interest.*

Man will have to create more homogeneous agricultural land to supply food for the hungry. He will have to apply pesticides and fertilizers to increase harvest efficiency. The need to increase population density and to develop industries in underdeveloped countries will increase pollution and harmful waste products. Ecosystems will have to be disturbed and destroyed to supply space for cities and industrial areas for the increasing population. These are irrefutable necessities. They are in direct conflict with the conscientious, responsible and engaged appeals of ecologists to minimize disturbances and to prevent irrevocable damages to our ecosystems. There are and there will be irrevocable changes. The need for quality of life is at odds with the need for bread. The recreational and spiritual value of undisturbed landscapes is in conflict with the value of land as a source of food and as space for houses and factories. The chances of reconciliation between the socially responsible desire to save ecosystems, and the social need to disturb and destroy ecosystems are diminished every year by the increasing density of the human population.

References

Barrett, G.W. 1969. The effects of an acute insecticide stress on a semienclosed grassland ecosystem. Ecology 49: 1019–1035.

Carlander, K. D. 1955. The standing crop of fish in lakes. J. Fish. Res. Board Canada 12: 543–570.

Cattaneo, A. and S. Ghittori. 1975. The development of benthonic phytocoenosis on artificial substrates in the Ticino River. Oecologia (in press).

Edmonson, W. T. 1972. The present condition of Lake Washington. Verh. Internat. Vereinig. Limnol. 18: 284–291.

Embree, D. G. 1966. The role of introduced parasites in the control of wintermoth in Nova Scotia. Canad. Entomol. 98: 1159–1168.

Goulden C. E. 1969. Developmental phases of the biocenosis. Proc. Nat. Acad. Sci. U.S. 62: 1066–1073.

Hassel, M. P. and G. C. Varley. 1969. New inductive population model for insect parasites and its bearing on biological control. Nature 223: 1133–1137.

Hurd, L. E. et al., 1971. Stability and diversity at three trophic levels in terrestrial successional ecosystems. Science 173: 1134–1136.

Johnston, D. W. and E. P. Odum. 1956. Breeding bird population in relation to plant succession on the Piedmont of Georgia. Ecology 37: 50–62.

Margalef, R. 1969. Diversity and stability: A practical proposal and model of interdependence. Brookhaven Symp. Biol. 22: 25–37.

Maurer, R. 1974. Die Vielfalt der Käfer- und Spinnenfauna des Wiesenbodens im Einflussbereich von Verkehrsimmissionen. Oecologia 14: 327–351.

Nuorteva, P. 1971. The synanthropy of birds as an expression of the ecological cycle disorder caused by urbanization. Ann. Zool. Fennici 8: 547–553.
Odum, E. P. et al., 1971. In: E. P. Odum, Fundamentals of ecology, 3rd ed. (Philadelphia), Table 6-2, p. 153.
Pielou, E. C. 1966. The measurement of diversity in different types of biological collections. J. theor. Biol. 13: 131–144.
Stockner, J. G. and W. W. Benson. 1967. The succession of diatom assemblages in the recent sediments of Lake Washington. Limnol. Oceanogr. 12: 513–352.
Watt, K. E. F. 1965. Community stability and the strategy of biological control. Canad. Entomol. 97: 887–895.
Watt, K. E. F. 1973. Principles of Environmental Science. McGraw-Hill, N.Y. 1973.
Weinberg, A. M. and R. P. Hammond. 1971. Global effects of increased use of energy. Proc. 4th Intern. Conf. Peaceful Uses of Atomic energy. Geneve 1971.
Wilhm, J. L. 1967. Comparison of some diversity indices applied to populations of benthic macroinvertebrates in a stream receiving organic wastes. J. Water Poll. Cont. Fed. 39: 1673–1683.
Wilson, E. O. 1969. The species equilibrium. Brookhaven Symp. Biol. 22: 38–47.

Author's address:

Jürgen Jacobs
Zoologisches Institut der Universität
8 München 2, Luisenstr. 14
W. Germany

On the vulnerability of ecosystems disturbed by man[1]

John Harte and Donald Levy

Abstract

The Liapunov stability theory is applied to models of energy and nutrient flow in ecosystems. The domain of stability under non-infinitesimal perturbations is discussed and significant differences are pointed out between models with and without detritus-decomposer feedback loops. Possible practical implications are suggested. Speculations concerning the role of fluctuations in ecosystems and the possibility of determining successional trends from an optimization procedure are also discussed.

Introduction

The ability to predict ecosystem instabilities is of great importance today because so many environmental conflicts are essentially disputes about stability—they boil down to a difference of opinion about whether a given man-induced disturbance of a system is likely to result in a severe disruption or merely a gentle recoil. What intensifies the importance of this problem is the fact that our species is now capable of adding to or subtracting from our natural surroundings on a scale comparable to the scale of natural processes. Thus, for example, the intensity at which waste heat is released by urban dwellers into their surroundings is in many cities 20% or more of the solar flux. Other disturbances such as the unsettling and dispersal of heavy metals into marine food chains during dredging operations, the disruption of fresh water supplies and sub-surface organisms during and after strip mining operations, or the release of toxic substances in fuel combustion, are not merely perceptible, but gross alterations of our no-longer natural environment. Potential instabilities that may result from these and other perturbations range from the loss of certain species to the creation of local dustbowls to global climate modification.

Because the disturbances of our environment are not infinitesimal but finite, the traditional tools for studying the stability of complex systems are clearly inadequate. From a practical view, not only do we have to deal with finite perturbations, but also with ecosystems which do not settle down to precisely their unperturbed states after we disturb them. Rather, what we can reasonably

[1] Research supported by the U.S. Atomic Energy Commission

hope for is that the initial perturbation will not propagate in such a way that the system is pushed beyond tolerable limits.

Thus, we are led to the concept of practical stability. This concept is intermediate between local and global stability. Local stability, utilizing the community matrix approach to a linearized system, is a mathematical nicety but, as we have mentioned, both too weak (as it is only reliable for infinitesimal perturbations) and too stringent (as it requires the system to return to its unperturbed state after the perturbation.) On the other hand, the requirement of global stability is too strong because we do not expect real systems to be stable under arbitrarily large perturbations.

A mathematical method exists for dealing with practical stability—a modification of the Liapunov Direct Method (Lasalle & Lefshetz, 1961). The results of some ecosystem studies using this method will be described here. We have obtained some interesting results, especially pertaining to the role of decomposers and feedback loops in an ecosystem. Furthermore, our methods allow some new insight into the role of fluctuations in systems and a possible understanding of the direction of successional trends.

The goal of our studies is the elucidation of ecosystem parameters which correlate with stability. To be of practical use, such quantities should depend upon general features of the system, such as the topology of the material and energy flow pathways, and not upon detailed knowledge of all the system's components and their dynamical interactions. Currently, much empirical activity in ecology is focused on measurements of quantities such as the biomasses and productivities of the components of the system, species diversity, and retention times of various nutrients. These measurements, while important, do not reflect the organization of an ecosystem and have not allowed ecologists to infer or understand ecosystem stability properties. Rather, they are largely indicators of the state of the components. If this work is to be successful, that is verifiable and of practical use, then it must ultimately point the way to measurable indicators of ecosystem organization and stability. A lot of work lies ahead.

Mathematical preliminaries

Assume we are given a reasonable mathematical model describing an ecosystem which can be written in the form of a set of coupled, non-linear, first order differential equations for the time rate of change of the components of the system. By 'component' we refer quite generally to the energy content, or the carbon content, or any other convenient measure (e.g. DDT content) of individuals or species or conveniently chosen aggregates of species or perhaps just physical sectors comprising the ecosystem. These equations are assumed to have the form

$$\frac{dx_i}{dt} = X_i(x_1, \ldots, x_N;\ w_1, \ldots, w_m) \qquad i = 1, \ldots, N \tag{1}$$

The x_i refer to the components of the system and the w_i are any other parameters upon which the time derivatives may depend. Phenomena such as time delays or stochasticity can be incorporated within this general form.

Suppose we are given an initial, unperturbed state of the system, $\bar{x}_i$, which may be time-independent (a steady state) or time-dependent (e.g. a limit cycle.) In Figure 1 we plot the trajectory of such a state. If the state is perturbed at some

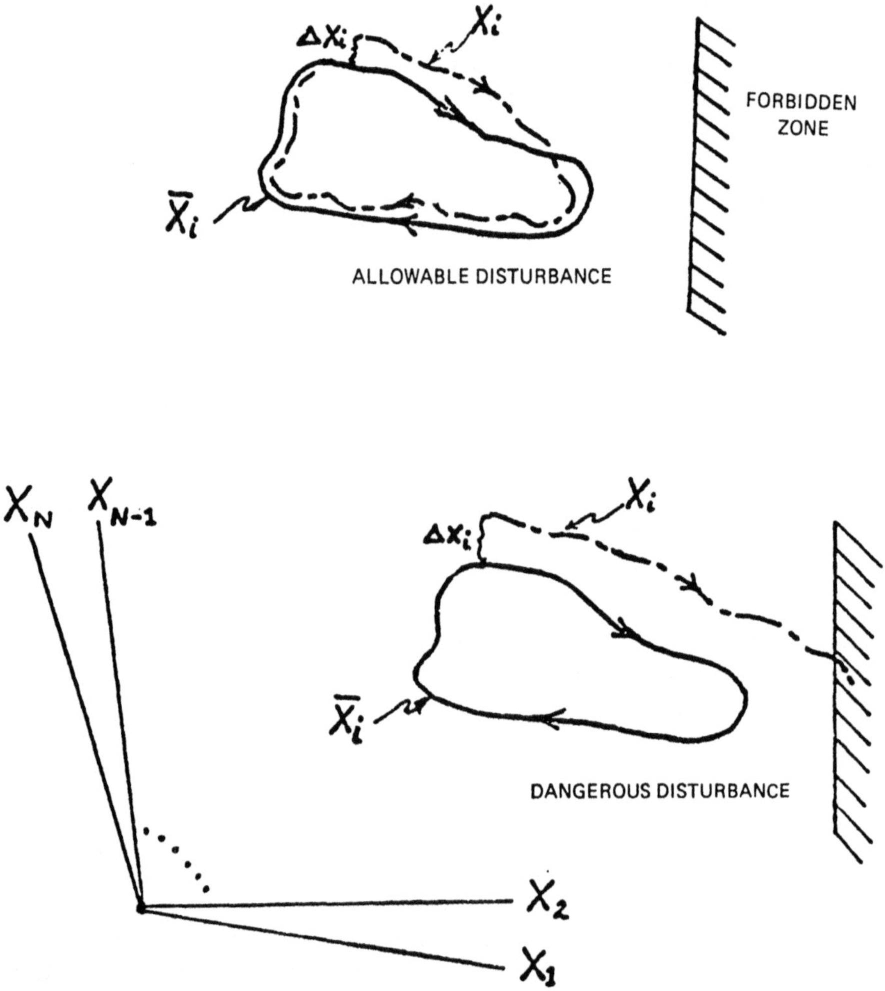

Figure 1. Time evolution of unperturbed and perturbed states. The axes label the component of the system. The solid line represents the unperturbed system ($\bar{x}_i$) and would be a single point for a steady state ($\bar{x}_i$ = constant). The dashed line represents the perturbed state and its subsequent time evolution. In the top diagram the perturbed state remains near the unperturbed state, while in the bottom diagram the perturbed state wanders into a forbidden region (e.g. a region where algae concentration accelerates dramatically).

time to a new value $\bar{x}_i + \Delta x_i$ shown in the figure (Δx_i not necessarily infinitesimal) then two options (see Fig. 1) are possible: (i) the perturbed state, x_i, may or may not return ultimately to the unperturbed state, $\bar{x}_i$, but it will never evolve further from the unperturbed state than some preassigned tolerance; (ii) it will evolve in time so as to exceed the preassigned tolerance. Our problem is to determine which option occurs.

The Appendix summarizes the Liapunov direct method for stability analysis. As shown, the crux of the method is to construct a function, $L(\Delta x_1, \ldots, \Delta x_N)$, which vanishes at the origin, and within some domain about the origin is positive and monotonically increasing with Δx_i, and in addition has a negative time derivative. For initial perturbations confined to that domain, the existence of such a function guarantees both the asymptotic stability of the system (i.e. the x_i will ultimately settle at $\bar{x}_i$) and a finite domain of practical stability.[1] The size of this domain of practical stability depends upon the preassigned tolerance. A slight modification[2] of the asymptotic stability criteria allows treatment of the more realistic case in which it is not required that the system return asymptotically to its unperturbed value. This, and other subtleties of the method such as the extension to the case in which the function L is explicitly time dependent or the case in which the perturbation is made not only upon the $\bar{x}_i$ but also upon the form of the equations of motion, are easily handled but will not be delved into now as they would only obscure the underlying principles which we wish to elucidate here.

For a wide class of ecological models,[3] some of which are described in the following section, a Liapunov function can be constructed for any initial steady state or periodic[4] state, $\bar{x}_i$. This function has the property that it vanishes at the origin, it is monotonically increasing in the entire Δx_i plane, and

$$\frac{-dL}{dt} = \sum_i \sum_j B_{ij}(x, \bar{x}, w)\, \Delta x_i\, \Delta x_j \tag{2}$$

[1] For practical stability, we require that at *finite times* the preassigned tolerance is not exceeded. Thus the domain of asymptotic stability could be larger than the domain of practical stability.

[2] See the discussion following p. 121 of Lasalle & Lefshetz (1961).

[3] This class of models *includes* those characterized as follows: Separate the net increasing and decreasing contributions to dx_i/dt by writing $dx_i/dt = f_i(x_1, \ldots, x_N) - g_i(x_1, \ldots, x_N)$ where f_i and g_i are positive and can be expanded in a sum of products of positive powers of the x_i. Further assume that $g_i(x_i = 0, x_{j \neq i} \text{ arbitrary}) = 0$, that f_i does not grow faster than linearly in x_i and that $f_i/g_i \xrightarrow[x_i \to \infty]{} 0$. Then such a B-matrix can be constructed. Of course, a wider class of models which are not expressible as sums of products of powers and which are quite difficult to characterize, will also lead to such a B-matrix.

[4] The treatment of an unperturbed periodic or nearly periodic state involves an averaging procedure which will be discussed in a forthcoming paper. For the rest of this paper, attention will be limited to steady states.

where the B_{ii} are strictly positive for all values of the x_i. Moreover, in realistic models, many of the B_{ij}, for $i \neq j$, are zero.

It is convenient to write the coefficients, B_{ij}, in the form of a symmetric matrix, B, hereafter called the B-matrix. The matrix elements are given by $[B_{ij} = \frac{1}{2}(B_{ij} + B_{ji})$. Now, a theorem on the positive definiteness of quadratic forms asserts that the form

$$q = \sum_i \sum_j Q_{ij} Y_i Y_j \tag{3}$$

is positive for all values of the Y_i if and only if the determinants of all the principal minors of the symmetric matrix of coefficients, Q, are positive (Fraser et al., 1957). Therefore, the domain of asymptotic stability of our system is at least as big[1] as the domain of the Δx_i for which the determinants of the principal minors of the matrix $B(x, \bar{x}, w)$ are positive. We emphasize that the B-matrix is not the community matrix. The latter describes infinitesimal stability of a linearized system while the B-matrix encapsulates the finite stability properties of a non-linear system.

Clearly, were it not for the presence of off-diagonal, non-zero elements in the B-matrix, we would have global stability. It is the organizational structure (patterns of pathways) of the ecosystem which determines which of the off-diagonal elements are non-zero, and therefore which places limits on the size of the domain of stability. In the following section we explore the implications of these ideas for various models of ecosystem.

Models and applications

Let us consider three broad categories of ecological organization:

1. *Open flow without cycling.* An example would be the flow of energy through the pathways of the food web from photosynthesizers on up to top carnivores. This is subsidized and therefore open flow, the source of sustenance being the sun. Admittedly a certain fraction of chemical energy is recoverable from detritus but it is usually a good approximation to ignore this.

[1] We say 'at least as big' because the actual domain of stability can be larger than that calculated from the principal minors. This is true for two reasons. First, the condition on the determinants arose from the requirement that the quadratic form be positive for all values of the Δx_i's. Yet the condition restricts the Δx_i's and thus the requirement on the quadratic form was overly stringent. Secondly, even if some of the determinants are negative so that dL/dt is no longer negative definite, dL/dt is not necessarily positive definite and thus there may not necessarily be a true instability. A better Liapunov function might be needed to resolve this ambiguity. For both these reasons we have a built in 'safety factor' in our analysis. We suspect that safety factors are desirable in practical ecosystem stability modeling if for no other reason than that model descriptions of ecosystems are inevitably only approximate. It remains to be seen whether this is the most appropriate way to build in the margin of safety.

2. *Closed flow with cycling.* An example would be any global material cycle for which the number of molecules of the material is conserved. A mathematical description of a closed cycle is, however, likely to be elusive because of the difficulty in accounting for all of the compartments in the cycle. Perhaps the global carbon cycle is the most natural one to model, with the dominant compartments being the atmosphere, plants, organic litter, decomposers, the oceans, animals, fossil fuel and the geosphere. In practice, most models will be geographically non-global and will not incorporate all compartments; thus one is led to:

3. *Open flow with cycling.* The nitrogen flow in a field is a fine example. Inputs and outputs such as the addition of fertilizer or washout from erosion might be driving forces behind this open flow, and yet the character and stability of the steady state or limit cycle solutions will be strongly influenced by the cycling capability of the system.

In Figures 2-4 we illustrate examples of these three types of organization. The pictures illustrate the pathway patterns. In addition, model equations are present which correspond to the flow diagrams. Other equations could be written—we have only shown these in order to focus on specific examples. What can we learn about the stability properties of these three systems from a Liapunov analysis, and, in particular, what properties are reasonably independent of the detailed mathematical model used to describe the flow diagrams?

Type i Systems. The Liapunov method has been employed by Huang and Morowitz (1972) to analyze the stability properties of the steady state solutions of the Lotka-Volterra equations for predator-prey interactions. These authors show that if the $\bar{x}_i$ are constant, then

$$L = \sum_i L_i = \sum_i \tau_i \left[x_i - \bar{x}_i - \bar{x}_i \ln \left(\frac{x_i}{\bar{x}_i} \right) \right] \tag{4}$$

is a Liapunov function for the system. Moreover

$$B = -\begin{pmatrix} k_{11} & & & & \bigcirc \\ & k_{22} & & & \\ & & \cdot & & \\ & & & \cdot & \\ \bigcirc & & & \cdot & k_{NN} \end{pmatrix} \tag{5}$$

for all x_i, indicating global stability if all $k_{ii} < 0$. Because the steady state solutions are globally stable, the equations clearly possess no limit cycle solutions. On the other hand, if we set the $k_{ii} = 0$ then the equations do possess oscillatory solutions but they are unfortunately not asymptotically stable nor are they structurally stable against small changes in the form of the equations of motion.

If a more general mathematical model describing the unidirectional flow of energy through an ecosystem is employed, restricted only by the constraints mentioned in footnote 3 (p. 211), eqn 4 is still a Liapunov function but the structure of the B-matrix is more complicated. In general, simply increasing the number of trophic levels will not affect stability, but increasing the number of pathways in the food web by introducing, for example, more competitors at each trophic level will add off-diagonal elements to the B-matrix; this tends to diminish the size of the domain of stability.

Type ii Systems. Figure 3 shows the pattern of pathways of a closed nutrient cycle. This cycle, and the model equations shown in the figure are a simple representation of carbon flow in a four level system consisting of photosynthesizers, the inorganic nutrient pool (which we take in this case to be the atmosphere), the decomposers, and organic litter (fallen leaves, dead trees, etc.) We have assumed that negligible amounts of carbon are added to or lost from the system (e.g. there is no exchange with the ocean.)

It is possible to construct a Liapunov function for this system and with it to establish the asymptotic stability of its steady state solutions under the class of

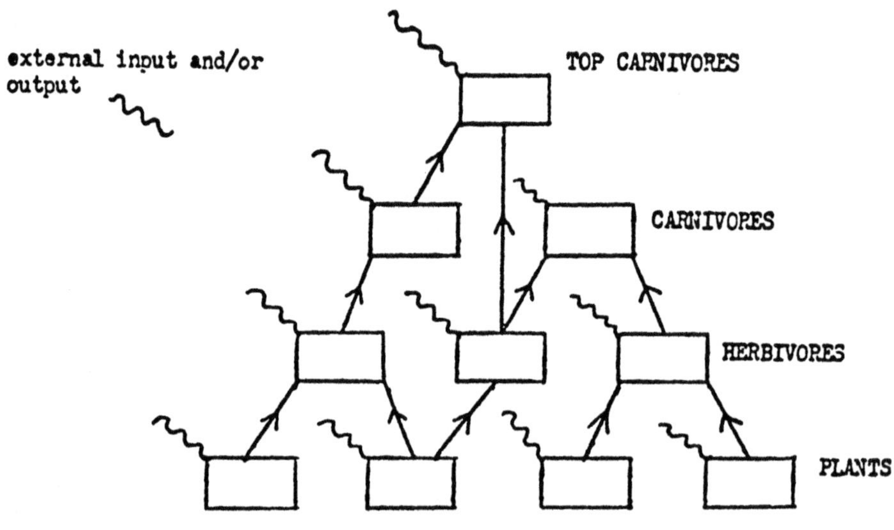

$$\frac{dX_i}{dt} = \alpha_i X_i + \sum_{j=1}^{N} \frac{k_{ij}}{\tau_i} X_i X_j$$

where $k_{ii}<0,\ k_{ij} = -k_{ji}\ (i \neq j)$

Figure 2. A schematic figure of a simple open system with no cycling of energy flow. At each trophic level above the plants, energy is lost. A typical set of equations describing such a system is the Lotka-Volterra equations shown in the figure. In these equations: the α_i are simple death rates, the k_{ii} are related to carrying capacities, the k_{ij} are the interaction terms, and the τ_i are retention factors. The simple form of the Lotka-Volterra equations and the antisymmetry constraints limit their usefulness.

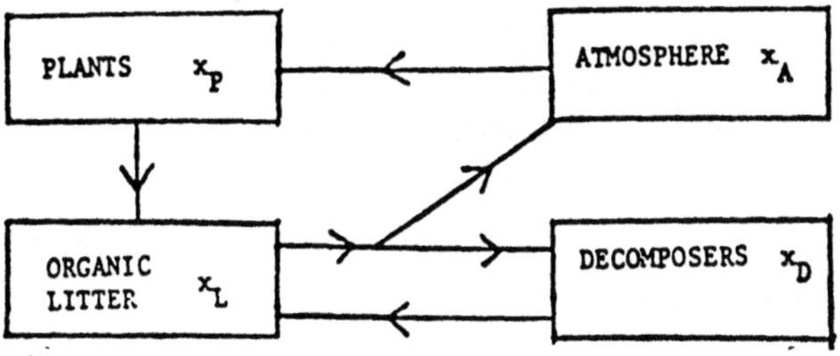

$$\frac{dX_P}{dt} = -\underbrace{\alpha_P X_P - \gamma_P X_P{}^2}_{1} + \underbrace{\beta_{PA} X_P X_A}_{2}$$

$$\frac{dX_A}{dt} = -\underbrace{\beta_{PA} X_P X_A}_{3} + \underbrace{\sigma_{DL} X_D X_L}_{4}$$

$$\frac{dX_D}{dt} = -\underbrace{\alpha_D X_D - \gamma_D X_D{}^2}_{5} + \underbrace{\beta_{DL} X_D X_L}_{6}$$

$$\frac{dX_L}{dt} = \underbrace{\alpha_P X_P + \gamma_P X_P{}^2 + \alpha_D X_D + \gamma_D X_D{}^2}_{7}$$

$$- \underbrace{(\beta_{DL} + \sigma_{DL})\, X_D X_L}_{8}$$

note the system is closed, i.e.

$$\sum_i \frac{dX_i}{dt} \equiv 0 \quad \text{or} \quad \sum_i X_i = \text{constant}$$

Figure 3. A simple closed system with carbon cycling consisting of plants, atmosphere (CO_2), decomposers, and organic litter. The x_i's measure carbon content. The various numbered terms in the equations indicate: (1) plant death rate including resource saturation effect; (2) plant growth due to CO_2 absorption during photosynthesis; (3) decrease in CO_2 due to plant absorption; (4) production of CO_2 by decomposer action on litter; (5) decomposer death rate including resource saturation effect; (6) decomposers growth due to feeding on litter; (7) litter increase due to decomposer and plant death; (8) litter decrease by decomposer action. (All constants are positive.)

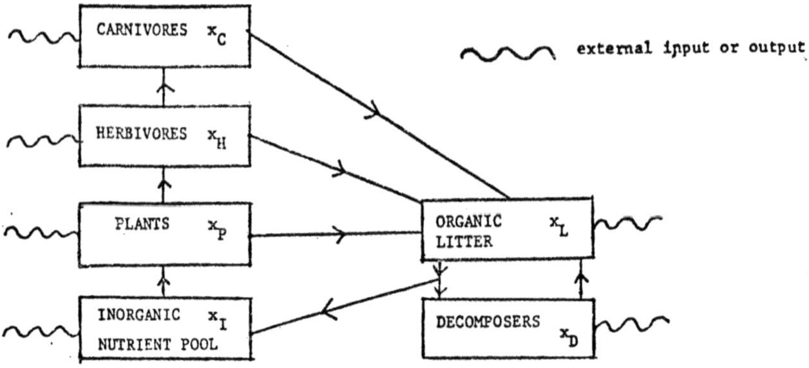

$$\frac{dX_C}{dt} = -\underbrace{\alpha_C X_C - \gamma_C X_C^2}_{1} + \underbrace{\beta_{CH} X_C X_H}_{2}$$

$$\frac{dX_H}{dt} = -\underbrace{\alpha_H X_H - \gamma_H X_H^2}_{3} - \underbrace{\beta_{HC} X_H X_C}_{4} + \underbrace{\beta_{HP} X_H X_P}_{5}$$

$$\frac{dX_P}{dt} = -\underbrace{\alpha_P X_P - \gamma_P X_P^2}_{6} - \underbrace{\beta_{PH} X_P X_H}_{7} + \underbrace{\beta_{PI} X_P X_I}_{8}$$

$$\frac{dX_I}{dt} = \underbrace{\Omega_I}_{9} - \underbrace{\alpha_I X_I}_{10} - \underbrace{\beta_{IP} X_I X_P}_{11} + \underbrace{\sigma_{DL} X_D X_L}_{12}$$

$$\frac{dX_D}{dt} = -\underbrace{\alpha_D X_D - \gamma_D X_D^2}_{13} + \underbrace{\beta_{DL} X_D X_L}_{14}$$

$$\frac{dX_L}{dt} = \underbrace{\Omega_L}_{15} - \underbrace{\alpha_L X_L}_{16} + \underbrace{\Gamma\{(\beta_{HC} - \beta_{CH}) X_C X_H + (\beta_{PH} - \beta_{HP}) X_P X_H\}}_{17}$$

$$\underbrace{-\beta_{LD} X_L X_D}_{18} + \underbrace{\Gamma\{\alpha_C X_C + \gamma_C X_C^2 + \alpha_H X_H + \gamma_H X_H^2 + \alpha_P X_P + \gamma_P X_P^2 + \alpha_D X_D + \gamma_D X_D^2\}}_{19}$$

Constraints are: $\beta_{HC} > \beta_{CH}$, $\beta_{PH} > \beta_{HP}$, $\beta_{LD} > \beta_{DL} + \sigma_{DL}$

Figure 4. A simple open system with nutrient cycling consisting of carnivores, herbivores, plants, inorganic nutrient pool, decomposers, and organic litter. The numbered terms indicate: (1) carnivore death rate including resource saturation effect; (2) carnivore growth due to eating herbivores; (3) herbivore death rate including resource saturation effect; (4) herbivore decrease due to carnivore grazing; (5) herbivore increase due to feeding on grass; (6) plant death rate including resource saturation effect; (7) plant decrease due to herbivore grazing; (8) plant growth due to absorbtion of inorganic nutrients; (9) addition of inorganic nutrient pool (e.g. fertilizer); (10) washout of inorganic nutrient; (11) nutrient decrease due to plant use; (12) nutrient increase due to decomposer action on litter; (13) decomposer death rate including resource saturation effect; (14) decomposer growth due to feeding on litter; (15) addition to litter (e.g. sewage dumping); (16) washout of litter; (17) litter increase from excrement (Γ is efficiency factor); (18) litter decrease due to decomposer action; (19) litter increase due to death of organisms. (All constants are positive.)

finite perturbations which are constrained to conserve the total amount of carbon. The Liapunov function is

$$L = \sum_i c_i(\bar{x}_i, \alpha, \beta, \sigma)\left[x_i - \bar{x}_i - \bar{x}_i \ln\left(\frac{x_i}{\bar{x}_i}\right)\right] \tag{6}$$

where the c_i are moderately complicated functions of the $\bar{x}_i$.

Thus the closed system is asymptotically stable against the arbitrary sliding of carbon from one level to another. From this result, and the fact that a steady state solution exists for every value of the total quantity of carbon in the system, it follows that this closed system is not asymptotically stable against perturbations which do not conserve the total amount of carbon. If the perturbation changes the total amount of carbon, then a new steady state will be approached asymptotically. An interesting question then arises: which steady state solutions are approached relatively rapidly when disturbed? This is amplified upon in Section 4 where we discuss successional trends.

It is straightforward but tedious to include more compartments such as the oceans into the model. We have not looked in detail at extremely complicated and inclusive models of the global carbon cycle, but from experience gained by working with relatively simple systems we suspect that the above results will remain valid for the quite general class of models characterized in footnote 3 provided the system is closed.

Type iii systems. If we open the system, that is allow for the incomplete cycling of the nutrient, then the situation changes. For example, consider the flow of a nutrient (such as nitrogen) in a six level system consisting of carnivores, herbivores, photosynthesizers, inorganic nutrient pool, decomposers and organic litter (plant litter, excrement and corpses) (Austin & Cook, 1979). Referring to Figure 4 we note that in our model equations we have adjoined a simple Lotka-Volterra-type predator-prey web upon the substratum of feedback dynamics describing the detritus—decomposer path ways.

A Liapunov function can be constructed, again of the general form of eqn 6. Its properties are best encapsulated by the B-matrix, which for steady state solutions has the form:

$$B = \begin{pmatrix} a_{CC} & 0 & 0 & 0 & 0 & -b_{CL} \\ 0 & a_{HH} & 0 & 0 & 0 & -b_{HL} \\ 0 & 0 & a_{PP} & 0 & 0 & -b_{PL} \\ 0 & 0 & 0 & a_{II} & -b_{ID} & -b_{IL} \\ 0 & 0 & 0 & -b_{ID} & a_{DD} & -b_{DL} \\ -b_{CL} & -b_{HL} & -b_{PL} & -b_{IL} & -b_{DL} & a_{LL} \end{pmatrix} \tag{7}$$

where the a_{ii} and b_{ij} are always positive. The a's and b's are simple functions of the x_i, $\bar{x}_i$ and the parameters c_i, α_i, Ω_i, γ_i, β_{ij}, σ_{ij} (see Fig. 4 for an explanation of the symbols). By the theorem on quadratic forms discussed above, our system is stable for perturbations which are initially within a domain of the x_i such that the determinant of the principal minors of B are positive. Now the first four principal minors are diagonal and clearly positive. Adding more superstructure to the system (i.e. more predator-prey links in the Lotka-Volterra part of the system) would not affect the positivity of the first $N - 2$ determinants.[1] Constraints on the domain of stability, if they exist, will show up in the evaluation of the last two determinants. The fifth determinant is given by

$$a_{CC}a_{HH}a_{PP}[a_{II}a_{DD} - b^2_{ID}]$$
$$= c_C\alpha_C c_H\alpha_H c_P\gamma_P\left[\frac{c_I c_D(\sigma x_D x_L + \Omega_I)}{x_I \bar{x}_I}\gamma_D - \frac{c_I^2\sigma^2(x_L + \bar{x}_L)^2}{16\,\bar{x}_I^2}\right] \tag{8}$$

We see that the domain of stability may now no longer be global; for fixed $\bar{x}_i$ and for sufficiently small values of Ω_i and x_D, or for sufficiently large values of x_I or x_L, the determinant becomes negative. Thus type iii systems can be quite vulnerable to perturbations in the litter, the inorganic nutrient pool or the decomposers.

The sixth determinant is complicated and we have not yet extracted all the information in it. For a range of cycling rate parameters (Γ and σ) and external input and output parameters (Ω_I, Ω_L, α_I, α_L) a finite domain of stability can be shown to exist. It is possible to show that in type iii systems in which the external input rate of litter is proportional to the amount of litter present, that a certain critical minimum value of the cycling efficiency parameter, Γ, is necessary in order to have a finite domain of asymptotic stability. On the other hand, if the system is approximately closed, i.e. external inputs and outputs relatively small, then the sixth determinant can become negative. This is simply a reflection of the fact discussed earlier that exactly closed systems are not asymptotically stable against displacements which do not conserve the total quantity of nutrient. This instability is a relatively harmless one, however, as long as the perturbed system does not evolve far away from where it is initially perturbed to. It remains to be seen whether thresholds of dangerous instability are more likely in high-Γ or low-Γ systems and in high-Ω_i or low-Ω_i systems. There are many other unanswered questions which we hope to explore. For example, do systems in which the inequality $\alpha_i\bar{x}_i \ll \gamma_i\bar{x}_i^2$ is satisfied tend to have a greater stability domain than systems satisfying the opposite inequality? In other words, are resource-limited systems more stable than those existing well below a saturation level?

An amusing relation between diversity and stability also emerges from this analysis. Let us enlarge the Lotka-Volterra 'superstructure' of the system by extending the matrix, eqn 7, to the upper left so that we consider an N-component

[1] Provided, of course, that the horizontal structure of the web does not grow so complex that the Lotka-Volterra form is impossible. The tendency will be for off-diagonal elements in the principal minors to diminish the domain of stability.

system. We denote by D_m^N the determinant of the mth principal minor of the $N \times N$ matrix and choose the values of x_D, x_L and x_I so that D_{N-1}^N is positive. How does D_N^N then behave as $N \to \infty$? The answer depends upon two factors: the shape of the trophic structure of the system and the ratios β_{ij}/β_{ji}. In general, D_N^N will remain positive as N increases and thus systems with an ever-increasing number of interacting components, arranged vertically in trophic hierarchy, will remain stable.

In order to show this we number the rows and columns of the $N \times N$ matrix in an unorthodox manner, letting N denote the first row or column and 1 the last. Thus the matrix element a_{NN} is that appearing in the top left corner of the matrix and corresponds to the top carnivore. Then the following recursion relations are easily derived:

$$D_{N-1}^N = a_{NN} D_{N-2}^{N-1}$$

$$D_N^N = a_{NN} D_{N-1}^{N-1} - \frac{b_{1N}^2}{a_{NN}} D_{N-1}^N \tag{9}$$

These can be easily solved. Let $\phi = D_2^3$ and $\psi = D_3^3$ both of which are assumed to be positive as is required for stability. Then

$$D_{N-1}^N = \phi \prod_{j=4}^{N} a_{jj}$$

$$D_N^N = \left(\psi - \phi \sum_{i=4}^{N} \frac{b_{1i}^2}{a_{ii}}\right) \prod_{j=4}^{N} a_{jj} \tag{10}$$

D_{N-1}^N clearly remains positive if the a_{ii}'s are positive (or γ_i positive in Fig. 4). The value of D_N^N will depend upon the b_{1i}'s and the a_{ii}'s. Now b_{1i} is proportional to the rate, per unit mass, at which x_i is cycled back to the organic litter level. We expect this quantity to be roughly independent of i and we henceforth take it to be a constant. Referring to eqn 10 we note that if a_{ii} increases less rapidly than i, then the summation will diverge as $N \to \infty$ and D_N^N will become negative at some critical value of N. D_N^N can remain positive as $N \to \infty$ only if a_{ii} increases faster than i.

What does this imply? For $i > 3$, $a_{ii} = c_i \gamma_i$. The γ_i behave roughly proportional to x_i^{-1} and thus increase with i for ordinary trophic hierarchies. Moreover

$$\frac{c_i}{c_{i-1}} = \frac{\beta_{i-1,i}}{\beta_{i,i-1}} \geq 1 \tag{11}$$

since retention is $\leq 100\%$ and so the c_i's will not decrease with i. If $\beta_{i,i-1}$ is a constant multiple of $\beta_{i-1,i}$ then $c_i = c^i$ and the sum will not diverge. To create a divergence and thus drive the determinant negative, we would have to assume that both $\beta_{i,i-1} \to \beta_{i-1,i}$ and $x_i \to x_{i-1}$ as $i \to \infty$. Thus the trophic structure would not peak as rapidly as is usually observed in real systems. Such unpeaked systems could exist but we have shown that they are likely to become unstable rapidly, as N grows large. Moreover, systems with inverted structure in which the top levels are more 'populated' than the lower ones should be of very simple (small N) structure for stability. For systems, with peaked trophic structure and small retention factors ($\beta_{i,i-1}/\beta_{i-1,1}$), increasing the value of N will not seriously affect stability.

Summary and speculations

We have described here several results of an investigation of the finite stability domain of ecosystem models including those incorporating decomposer and detritus pathways. While only the surface of this subject has been scratched by our work, several pertinent results have emerged. Among these are two which may be of practical interest:

1. *Stability and diversity.* We have distinguished several kinds of diversity here. There is vertical diversity referring to the number of levels in the trophic structure, and horizontal diversity, referring to the variety of competitors at each level. And then there is diversity of species and diversity of pathways. What we have shown here is that increasing the number of trophic levels generally has no effect on the size of the domain of asymptotic stability. The exception to this occurs if the food web is not pyramidal in shape but rectangular or inverted. Then the system can rapidly destabilize as the number of levels grows. We have also shown that increasing horizontal diversity generally leads to a decreasing domain of asymptotic stability, although if the ratio of the number of pathways to the number of species is kept sufficiently bounded, then both can increase without diminishing stability.

2. *Sensitivity of feedback systems.* We have shown that damage to the decomposers or the organic or inorganic nutrient pools in an ecosystem is a potential source of instability—greater, perhaps, than that arising from tampering with the more visible predator-prey components of the system. Activities of man which diminish the cycling capability of an ecosystem should be viewed with caution if these results stand up under further analysis.

There are numerous practical problems to which stability analysis such as this might be applicable. Study of the global carbon cycle might reveal thresholds for climatic instability, or at least provide insight into the ultimate fate of the carbon dioxide released by fossil fuel consumption.

Our methods might also be useful for evaluating the potential for rehabilitating strip-mined lands which have had their detritus-decomposer pathways altered. Insight into the vulnerability of desert and tundra systems with low reserves of litter and slow cycling times might also be obtainable.

We close with several speculations. Let us recall the observation that during the course of ecological succession, certain observables such as cycling rates, productivities, and biomass tend to show systematic time development, see for example Odum, 1969, p. 262. Is it possible to view succession as the progression of the system into an ever more resilient configuration? If so, then the evaluation of a suitable measure of resilience may provide a guide to the direction of these successional trends.

As a suitable measure, we propose the use of either the quantity

$$\Lambda = -\text{ minimum over } \Delta x_i \text{ of } \left(\frac{1}{2}\frac{d}{dt}\ln L\right) \tag{12}$$

or, if one is only concerned with neighborhood stability,

$$\Lambda = -\text{ minimum over } i \text{ of } (\lambda_i) \tag{13}$$

where the λ_i are the eigenvalues of the community matrix of the system. For a stable system, either is roughly a measure of the lowest resilience or recovery rate for a perturbed state to return to its unperturbed value. In the limit of small Δx_i, the two definitions of Λ usually agree. Λ is a quantity which should be of practical interest to those concerned with environmental impacts. For even though mathematical modeling may suggest that a system is asymptotically stable, a high resilience is still desirable as it proves a safety factor against the unexpected.

If we assume that Λ is maximized during succession, then we may be able to understand the course of succession. Moreover, if the equations of motion were sufficiently reliable, then if Λ is evaluated for a time-dependent solution it may be possible to use the dynamical equations to show that Λ *is* increasing in time.

We have only been able to apply this idea so far to several simple models. A number of simple two and three level systems describing the flow of carbon have been analyzed and lead to the result that Λ is maximized for a certain fixed ratio (which turns out to be 4) of the equilibrum amount of carbon in plants to carbon in the atmosphere. This result will be described in detail in a forthcoming paper. It will be interesting to determine how Λ depends upon such quantities as the total biomass, pathway diversity, or productivity of model and laboratory systems.

We have also begun to assess the role of fluctuations or noise in ecosystems. Several authors (see May, 1973 and references therein) have shown that fluctuations in the k_{ii} terms in Figure 2 are destabilizing. On the other hand we have obtained some preliminary evidence from computer generated solutions of our model equations that noise in the values of the k_{ij}, for $i \neq j$, renders the system more resilient—more like a piece of rubber than a crystal. To be more precise, consider the admittedly over-simplified Lotka-Volterra equations as an example. If the saturation effect is ignored ($k_{ii} = 0$) then the equations possess solutions with interesting cyclic time dependence, but this system is structurally unstable and the solutions possess no domain of asymptotic stability; hence the model is unrealistic. On the other hand, if the damping factors, k_{ii}, are negative definite, then the system is structurally stable, but the solutions all approach steady states. We hypothesize that in the latter case, with damping, the presence of small fluctuations in the k_{ij} (for $i \neq j$) will not only preserve the stability of the system but also excite the cyclic modes of the undamped system. A linear system could not obey this hypothesis; a more thorough study of non-linear systems is needed before this and other potential surprises are understood.

Much work remains in this exciting field. The search for a deeper understanding of the workings of complex, non-linear, self-organizing systems is a challenge in its own right; moreover, the results may be of ultimate benefit to our species.

Acknowledgements

We are extremely grateful to Michael Dudzik for sharing his insights and computational expertise. One of us (J. H.) is deeply appreciative of Harold Morowitz for his guidance and encouragement, and Daniel Botkin and Matthew Sobel for helpful conversations.

Appendix

The Liapunov direct method of stability analysis

Suppose you have a complicated non-linear model which describes how some system works. You want to investigate the stability properties of the solutions to the model equations, but you cannot actually solve the equations. The Liapunov Direct Method is then the appropriate tool to use, for without knowing explicit solutions, but only knowing the form of the dynamical equations, it can provide useful stability information about the system.

Application of the method proceeds as follows. The first step is to construct a function of the variables, Δx_i, which are the deviations of the system variables from their unperturbed values. This function must vanish when the Δx_i all vanish and it must increase from zero as any or all of the Δx_i become non-zero. That is, the function must be cup-shaped in some domain about $\Delta x_i = 0$. This first step is easy; many functions, the simplest of which is $L = \Delta x_1^2 + \Delta x_2^2 + \cdots + \Delta x_N^2$, will satisfy the conditions. The next step is to evaluate the time rate of change of the function, dL/dt. This is to be done using the equations of motion for the dx_i/dt ($= d\,\Delta x_i/dt$ if the unperturbed state is static), and the rule:

$$\frac{dL}{dt} = \sum_i \frac{\partial L}{\partial \Delta x_i}\frac{d\,\Delta x_i}{dt}$$

The last step is to examine the sign of dL/dt. If dL/dt is zero, the solutions are neutrally stable, i.e. if displaced from equlibrium, the system will neither return to its unperturbed value, nor will it wander far from it; it will simply remain in a displaced orbit. If dL/dt is negative in some domain about $\Delta x_i = 0$, then displacements of the system which are initially confined to within that domain will damp out and the system will return to its unperturbed value. Such a system is called 'asymptotically stable' and the range of perturbations which damp out is called the 'domain of asymptotic stability'. If dL/dt is positive, then the perturbations will grow in time and the system is unstable.

This deceptively simple result has one difficulty—in general, if you pick a function satisfying the two prescribed conditions and then evaluate its time derivative, you will often find that dL/dt does not have a single sign in some domain about $\Delta x_i = 0$; rather it will be positive in some directions and negative in others.

Then you learn nothing and have to try again with another function until a stability indicator is found. A theorem has been proven which guarantees that for systems with well defined stability properties, such a function indicating stability must exist. Unfortunately, for general systems, no algorithm exists for finding it. For conservative or dissipative mechanical systems, the Hamiltonian with the damping term neglected is often the appropriate stability indicator; for the equations of chemical kinetics, the Gibbs free energy often works; and in ecology, the stability indicators described in this paper appear to be of practical use.

It should be pointed out, however, that for ecological systems the functions we discuss here may not be the optimal ones. For open systems with nutrient cycling, our function indicates asymptotic stability only if the magnitude of the initial perturbation is not too large. A better function might indicate stability in larger domain. We urge interested readers to search for Liapunov functions for their favorite models. Trial and error techniques will be required at first, but we suspect that intuition and insight will be acquired in the process and hopefully our results here can be improved upon.

References

Austin, M. and B. Cook. 1974. Ecosystem stability: a result from an abstract analysis. J. Theor. Biol. 45: 435–458.

Fraser, R., W. Duncan and A. Collar. 1957. Elementary matrices. Cambridge University Press, Cambridge.

Huang, H. and H. Morowitz. 1972. A method for phenomenological analysis of ecological data. J. Theor. Biol. 35: 389.

Lasalle, J. and S. Lefshetz. 1961. Stability by Liapunov's direct method. Academic Press.

May, R., 1973. Stability and complexity in model ecosystems. Princeton University Press, Princeton, N.J.

Odum, E. P. 1969. The strategy of ecosystem development. Science 164: 262.

Authors' address:

John Harte
Lawrence Berkeley Laboratory
Energy & Environment Division
University of California
Berkeley, California 94720
U.S.A.

Donald Levy
Lawrence Berkeley Laboratory
Energy & Environment Division
University of California
Berkeley, California 94720
U.S.A.

Response of natural microbial communities to human activities

M. Alexander

Introduction

At the outset of my discussion, I wish to stress two crucial points. First, the quantitative contribution of microbial processes to the function of terrestrial, freshwater and marine ecosystems is almost totally unknown. Second, little compelling evidence exists showing that human activities are causing severe disturbances in major natural cycles or other geochemical processes of global concern that are catalyzed by microorganisms. On the other hand, little or no compelling evidence exists denying that such perturbations or significant upsets are occurring now or will occur if present trends are maintained. To a large extent, the second point arises because of our lack of a useful body of data on the first. Nevertheless, a reasonably large number of studies have been conducted, and a few generalizations, although they are still first approximations, can be advanced.

Contributing to our state of ignorance of what might in fact be a group of very serious problems are both microbiologists and ecologists. Microbiologists, including many of those who consider themselves to be environmental scientists, are largely concerned with in vitro problems. Typically, they examine one population at a time, and frequently they investigate not processes of ecological importance but rather those which are either of biochemical or genetic significance or which are fascinating because of their esoteric character. Thus, an enormous literature is available on the microbial metabolism of compounds of intracellular but not environmental importance, and a vast amount of genetic information exists on traits whose phenotypic expressions are not involved in helping to solve problems arising in aquatic or terrestrial ecosystems. By contrast, most ecologists who are not directly studying microbial processes view microorganisms merely as a resource, the so-called 'decomposers', that can conceptually be placed into a small black box to be treated much as an engineer might deal with a sewage treatment system, rather than as a biologist might look at a discrete community. The ecological literature thus shows a woeful lack of information which would allow microorganisms to be put in their rightful place in the function of ecosystems, instead they are merely relegated to the role of decomposing a variety of unspecified organic materials that results in a continued regeneration of nutrients required by plants on land and water and that leads to the removal from the environment of potentially harmful compounds.

Modern microbial ecologists are endeavoring to set right this sad situation. They are attempting to avoid the pitfalls commonly plaguing the pure culture microbiologists, even those who nominally work on topics of environmental concern but do so without a proper perspective of the interactions among populations in the community. Members of the new breed of microbial ecologists also seek to avoid the generalizations and platitudes so commonly expressed by our colleagues whose field of endeavor is the ecology of higher plants and animals. Natural microfloras are not merely the incinerators of nature, the so-called 'decomposers', and they are not simply laboratory playthings for the biochemist or the geneticist. Nevertheless, defining their role in nature in meaningful terms will require considerable time and effort.

Activities of microorganisms in natural ecosystems

What is meant by the term 'microorganism?' The bacteria, fungi and actinomycetes of terrestrial, freshwater and marine habitats that are responsible for a variety of critical processes are well known members of the microbial realm. In addition, however, the algae concerned with primary production in aquatic ecosystems are also included in the scope of microbiology, although admittedly some of the marine algae are not really microscopic in size. Protozoa are also within our area of concern, and their function in predation still requires more precise delineation. Finally, viruses are, of course, common subjects for research, but their role in natural ecosystems is obscure and probably minimal except as agents of disease.

The restrictions of time permit neither a discussion of the role of microorganisms in the function of ecosystems nor an evaluation of the limited quantitative data on the extent or rates of the various processes. Yet, it would probably serve a useful function to list what microorganisms are apparently doing in nature in order to allow for an assessment of the response of microbial communities to human activites. The algae inhabitating inland bodies of water and the oceans are major agents of photosynthesis, and they bring about much of the primary production in aquatic environments. Heterotrophic bacteria and fungi are essential in habitats also supporting higher plants inasmuch as they are responsible for the degradation of probably all naturally occurring organic molecules and many synthetic organic compounds and their conversion to inorganic products. These organisms also prevent the accumulation of potentially toxic intermediates which are known or are assumed to be formed in the pathway of breakdown of both natural and synthetic substances. These various processes lead to mineralization of the organic carbon and regeneration of this nutrient element in a form available to higher plants and aquatic algae, namely as CO_2. Mineralization and the consequent nutrient regeneration are transformations effected by microorganisms on, so far as is known, every element required for the synthesis of plant and animal constituents. Thus, microorganisms attack a multitude of compounds containing nitrogen, sulfur, phosphorus and other elements and convert them to the inorganic state so that plants can utilize the nutrient elements once again. These are the so-called 'decomposer' functions. At the same time, microorganisms assimilate the same elements, thereby rendering them unavailable to higher plants with occasional detrimental consequences.

Natural microfloras are also critical for other biogeochemical transformations. Thus, much of the O_2 that is liberated to the atmosphere results from activities of marine algae. Conversely, because bacteria and fungi probably are chiefly responsible for organic matter turnover, the same organisms must be major consumers of the O_2 in the atmosphere, converting that element into non-gaseous forms. Bacteria and algae, either free-living or in symbiotic associations, are the sole agents responsible for biological N_2 fixation, and thus they provide the biosphere with the element that often limits food production. Bacteria are responsible for probably nearly all of the nitrate formed in terrestrial and aquatic environments, and sulfate formation and sulfate reduction are likewise consequences largely or solely of microbial metabolism. The oxidation and reduction of sulfur compounds as well as the reduction of iron have enormous geochemical significance, and not only are compounds of these elements subject to microbial change but so too are compounds of a variety of other elements which react with the products of sulfur and iron metabolism (Alexander, 1971). The formation of humus in soil and sediments and its decomposition have been attributed to fungi and bacteria, and members of the same groups of organisms are undoubtedly largely responsible for the formation and maintenance of soil structure, which is so important to the normal development of rooted plants. Recent evidence, moreover, indicates that much of the air pollution of current concern, apart from that present in cities, is generated by microbial communities, and probably carbon monoxide and ethylene are almost wholly destroyed by these communities. Finally, microorganisms are probably responsible for synthesis of the growth factors required by many species of the phytoplankton, and they also eliminate the countless human, animal and plant pathogens continually discharged on the land and into waterways.

The influence of pollutants on microbial activities

Microorganisms participate in, or they are the sole agents for, an array of processes, and the number of possible ways by which human activities can upset these communities is consequently quite large. Yet, on reviewing the published research to ascertain how these various processes, activities and transformations are modified by man, one finds little quantitative information. The scarcity of data, however, has unfortunately not prevented fanciful and often unwarranted speculation about impending doom because of the presumed cessation of some crucial biogeochemical reaction sequence.

Because of the long recognized role of microorganisms in maintaining soil fertility, scientists concerned with agriculture and food production have evaluated some of the stresses associated with farm-related activities. In addition, the growing awareness of the importance of microbial communities to the maintenance of environmental quality has prompted the initiation of studies directed toward understanding whether the newer environmental pollutants inhibit one or another of the several processes. Yet, despite the multitude of biochemical changes and the enormous diversity of microorganisms, the effects of man-imposed stresses on only a few transformations and species have been evaluated to date. For example, although numerous estimates have been made of the influence of pesticides on the bacteria or occasionally fungi appearing on dilution plates prepared from soils, often there is little or no evidence that these particular species are of appreciable importance in the original habitat. This results from the fact that many microorganisms probably exist in nature in resting rather than in metabolically active stages. The effect of the same stress on the populations of major importance frequently has not been established, often because populations responsible for the major transformations have not been clearly delineated. Furthermore, such assessments of effects on numbers of organisms often have little meaning for understanding community function, because data showing a certain degree of inhibition of total numbers of organisms cannot be interpreted readily in terms of a certain extent of inhibition of some transformation that occurs in the habitat.

Research methods and results

Most of the assessments of man-imposed stresses in aquatic communities have been made using photosynthesis as the measure, either by estimating the rate of CO_2 fixation or of O_2 evolution. The photosynthetic algae, needless to say, are extremely important, and testing their response has the virtue of giving a remarkably sensitive assay for the influence of perturbations or chemicals inadvertently or deliberately introduced into inland or marine waters. On the other hand, the influence of these stresses on heterotrophic populations and on autotrophic bacteria in natural waters has largely been ignored.

In terrestrial habitats, the response to a stress is often evaluated by measuring the rate of carbon mineralization or the rate of O_2 consumption resulting from microbial respiration. The conversion of organic nitrogen compounds to ammonium and nitrate is also frequently used as a basis for concluding whether a community is under stress, but both nitrogen mineralization and respiration are brought about by a large number of dissimilar heterotrophs, so that marked inhibition of a group of sensitive organisms may lead to their replacement by more resistant organisms; as a result, the rate of the process may not change appreciably despite the suppression or even elimination of a major group of organisms. By contrast, nitrification is a remarkably sensitive process, and a number of chemicals have been found to inhibit this reaction sequence, one brought about by a remarkably small number of species of autotrophic bacteria.

Apart from measurements of these few reactions, the potential modification of the activities of terrestrial communities has received little attention.

The restraints of time likewise prevent any survey of the results that have been obtained to date, and no detailed review has yet been written on the effects of man's actions in general on microbial processes or communities, whether the actions be associated with agriculture, industry or other of society's activities. However, many individual reports or occasional reviews exist on the influence of pesticides on diverse categories of organisms (Alexander, 1969), SO_2 on lichens (Ferry et al., 1973), crude oil on algal primary productivity (Dickman, 1971), polychlorinated biphenyl on marine phytoplankton communities (Moore & Harriss, 1972), acid mine wastes on protozoa and algae (Lackey, 1939), mercury on phytoplankton species (Nuzzi, 1972), chlorination on photosynthesis and respiration by aquatic organisms (Brook & Baker, 1972), and heavy metals on marine and freshwater residents (Rice et al., 1973; Ruthven & Cairns, 1973).

The area of human activities which has been explored most thoroughly in terms of impact on microorganisms is agriculture, and the disturbances arising from cultivation, fertilization, the plowing under of crop residues and the use of pesticides have been well documented. There has also been a recent upsurge of interest in the impact of industrial effluents on the aquatic microbiota, but the body of data is extremely meager, especially in view of the multitude of compounds that are discharged into waterways, some of which are probably present in sufficient quantities to upset the biota modestly or seriously. The impact of pesticides, polychlorinated biphenyls, soil-derived nutrients and acidic mine wastes on populations in freshwater and marine habitats has also been the subject of some scrutiny, the extent of disturbance varying from none to catastrophic. Surprisingly, the possible influence of air pollutants on microbial communities has been almost totally neglected, except for the effect of SO_2 on lichens. It is likely that most air pollutants do not greatly inhibit heterotrophs because of their low concentrations and their rapid destruction by chemical or biological means in soils and waters.

Conclusion

Considering the available information, meager though it is, and bearing in mind the physiological versatility and rapid growth of microorganisms and the high species diversities of many communities, I do not feel there is any basis for concluding that we are approaching a cataclysm or that serious upsets of global concern are imminent, although ample data exist to show serious but local ecological disturbances. Conversely, I see no basis for a sense of complacency either. My chief feeling is neither one of fear nor of complacency but rather one of ignorance. I believe we just do not have adequate information to decide if there are or are not major or global disturbances resulting from the actions of society.

Owing to the scanty literature and the key role of microscopic organisms in ecosystem function and in the maintenance of human environments of acceptable quality, further work is therefore required to define more adequately how the actions of society alter microbial populations and communities and the biochemical processes they bring about.

References

Alexander, M. 1969. Microbial degradation and biological effects of pesticides in soil. In Soil biology: reviews of research. UNESCO, Paris. pp. 209–240.

Alexander, M. 1971. Microbial ecology. Wiley, New York.

Brook, A. J. and A. L. Baker. 1972. Chlorination at power plants: impact on phytoplankton productivity. Science 176: 1414–1415.

Dickman, M. 1971. Preliminary notes on changes in algal primary productivity following exposure to crude oil in the Canadian arctic. Can. Field. Nat. 85: 249–251.

Ferry, B. W., M. S. Baddely and D. L. Hawksworth. eds. 1973. Air pollution and lichens. Athlone Press, London.

Lackey, J. B. 1939. Aquatic life in waters polluted by acid mine waste. Public Health Reports Washington 54: 740–746.

Moore, S. A. and R. C. Harriss. 1972. Effects of polychlorinated biphenyl on marine phytoplankton communities. Nature, London 240: 356–358.

Nuzzi, R. 1972. Toxicity of mercury to phytoplankton. Nature, London. 237: 38–40.

Rice, H. V., D. A. Leighty and G. C. McLeod. 1973. The effects of some trace metals on marine phytoplankton. CRC Crit. Rev. Microbiol. 3: 27–49.

Ruthven, J. A. and J. Cairns. 1973. Response of fresh-water protozoan artificial communities to metals. J. Protozool. 20: 127–135.

Author's address:

M. Alexander
Laboratory of Soil Microbiology
Department of Agronomy
Cornell University
Ithaca, New York 14853
U.S.A.

A short comment

C. O. Tamm

We have heard it stated that man is part of the ecosystem and that most human perturbations have natural correspondences, even if man's activity may lead to more serious consequences because of the scale or frequency of the manipulations.

The uniqueness of man as a biological species is his ability to transfer information, not only between individuals but also over large distances and between generations. It is the special task of ecologists to transfer information to other people about what will happen following various human perturbations.

I would here like to point at certain ecological specialities in some human activities. Man has to harvest products of nature to get food and other useful materials. Some of us heard yesterday from Dr. Orians that 'generalists' among consumers had a lower rate of digestion than animals eating the same species but specialised on only one plant species. Man is certainly a 'generalist' and consequently he has a moderate ability to digest various plant materials. Therefore man has to select his food. Now it happens that almost all important food plants must be called pioneer plants from an ecological viewpoint. Their ancestors were often weeds or shore plants, which have been further bred for their ability to produce valuable seeds or storage organs, but not for their ability to compete in dense vegetation.

This means that in order to cultivate food plants man has usually to change the ecosystem in a direction favourable for pioneer plants with all the consequences this has, increasing erosion risks and leaching of nutrients. However, in agriculture in the wide sense there are some cases where man has apparently succeeded in establishing ecosystems stable over a long time under continuous human influence. The cultivation of paddy rice in eastern Asia is probably the best example. We may also point at the use of leguminous plants as a source of nitrogen instead of fertiliser, particularly in the case of grazing land. It seems to be possible to create fairly stable conditions in this way.

It might be imagined that forestry does not need to change the ecosystem towards pioneer stages, as many valuable trees are long-lived and at the present time are harvested from more or less virgin forests in various parts of the world. However, in a sustained yield type of forest management, where we also require high or at least reasonable yield, it seems necessary to harvest the trees as soon as they have attained suitable sizes. It is commonly believed that the net primary production of a forest ecosystem first increases with age until about stand closure and then decreases. There are two reasons for this decrease, one is the increase in non-photosynthetic tissue which causes higher respiration losses as stems and branches grow thicker with age. The other reason is that in many forest ecosystems a large part of the site's nutrients is eventually tied up in the stand in what we call biomass but which is actually dead wood and bark. A high yield presupposes that the trees are harvested before this decrease in production is too high. In addition, many desirable tree species, particularly many fast growing conifers, are trees of a pioneer type. Yet the change in direction to favouring pioneer stages in forest management is much less drastic than in conventional agriculture.

In the case of inadvertent consequences of human activities there is also often a question of change of rate of a natural development rather than the start of an entirely new development. Pollution of lakes is often compared with an increased ageing of lakes. Irrigation may lead to salinization in a dry climate, a process which also may occur naturally. The consequence of acid rain due to atmospheric pollution is increased leaching of the soil, not much different from the natural podsolization process. However, we also have phenomena outside the normal ecological experience, for instance pollution with heavy metals or other poisons which are not known in nature.

It is very important that we try to collect all ecological experience which can be applied to cases where we have some relevant information, either from earlier studied perturbations or from entirely natural processes. In the cases where we do not have such experiences, the advice of the ecologist to the decision-maker should be extreme caution and a request for resources for intensive research.

I would also like to emphasize one type of diversity which has not been much mentioned in the earlier discussions. This is the diversity in landscape, the mosaic of various ecosystems characteristic of many regions. Even if we as biologists regret every loss in biological diversity within ecosystems, it is quite as important to preserve the diversity of landscape as the diversity within each of the ecosystems constituting the landscape. In addition to the aesthetic values this may also serve the purpose of preserving animal species which need more than one biotope. I think we have reasons to include man among these animals.

Author's address:

C. O. Tamm
The Royal College of Forestry
S-104 05 Stockholm
Sweden

Discussion

Summarized by L. Vlijm and G. E. Likens

Participants: the authors M. Alexander (U.S.A.), J. Harte (U.S.A.), J. Jacobs (W. Germany), D. Levy (U.S.A.), C. O. Tamm (Sweden), the chairmen G. E. Likens (U.S.A.), L. Vlijm (The Netherlands), together with W. F. Blair (U.S.A.), E. D. Le Cren (U.K.), K. Curry-Lindahl (Kenya), M. J. Dunbar (Canada), M. Evenari (Israel), A. D. Hasler (U.S.A.), S. Kohlemainen (Puerto Rico), J. Kvet (Czechoslovakia), J. Mukiibi (Uganda), D. Müller-Dombois (U.S.A.), Z. Naveh (Israel), N. Polunin (Switzerland), L. J. Post (Canada), D. W. Schindler (Canada), R. Scossiroli (Italy), O. L. J. Vanderborght (Belgium).

The discussion, as reported here, is not in chronological order, as questions or remarks on various points have been grouped together by the panel. Several contributions were of more or less general importance. Some gave additional information to one of the papers, or criticized details.

The Chairman (*Likens*) opened the discussion by stating that given the need to harvest (water, minerals, timber, food, fish, fossil fuels) and management by man, which cause disturbances of natural ecosystems, a number of questions can be asked: (1) what is a 'natural system', and what are the boundaries? (2) What is the effect of man's activities on ecosystem functioning (observable and quantitative), with relation to (a) its biochemistry (leakage versus conservation of nutrients and water), and (b) its metabolism (such as changes in diversity influencing function). And (3) concerning the success of man in replacing 'normal' functions of natural ecosystems, (a) can he do so, and (b) should he do so?

It is clear that virtually no ecosystem is left that is not influenced by human activities. In the past such activities were thought to be an 'outside factor' as seen from such a system. Nowadays (as Jacob's paper showed) there is an increased understanding that man's activities should be built into the concept of ecosystem. Man is part of the natural world, and full understanding of his management and its consequences cannot be omitted (for example the spread of DDT all over the world).

The concept of 'ecosystem' is used with a variety of meanings, and great care is needed when formulating a concept which allows a variety of interpretations. (*Polunin*).

A basic understanding should be possible by changing from the 'how' attitude to the 'why' attitude. Perhaps this could help to overcome the sometimes metaphysically orientated, discussion of terminology, and allow us to reach a better predictive potential to inform policy-makers about their goals. (*Vanderborght*).

This is however only possible when data are available. It is obvious that in several fields (such as microbiology) more data are lacking than are known. Therefore, perhaps, the 'what' attitude should be stressed. (*Alexander*).

In relation to the position of man, it is clear that concepts of diversity and stability can only be used as operational terms when the pertaining system can be described (cf Jacobs); until then no generalising conclusions can be formulated. (*Müller-Dombois*).

Man's management leads to the conclusion that in the past 'natural ecosystems' provided a variety of 'services' free to man: clean water, clean air, recreation and aesthetics. At present these 'services' are used in such a way and on such a scale that disturbances occur. The free services disturbed by man's activities (agriculture, mining, forestry, fisheries etc) must then be put onto a direct cost-basis, and they usually require energy. It should be kept in mind that about 40% of fossil energy, as used by mankind, is used by the industrial system itself in providing energy. How energy payments can help to provide a suitable environment for the future is not at present clear. (*Likens*).

Do the data support the message of doom of many ecologists? If human population increase and the continued destruction by man of the world's ecosystems are inevitable, we might as well start writing an epitaph for the human species. (*Blair*).

They may not be inevitable. The point is that we do not know a dead certain and workable way to make them avoidable. (*Jacobs*).

The attention drawn to the 'overdominance' of man led to the question as to how far ecologists are able to optimise diversity to achieve secondary stability. Moral consideration of man's actions *vis-a-vis* other organisms, and their right to existence, should then be taken into account. (*Polunin*).

On the other hand, when man's actions are considered, one wonders whether 'natural ecosystems' should be conserved or 'man-made systems' created to fulfil the need to support life. (*Naveh*).

It is clear that the answers will have to be both complex and specific at the same time. (*Jacobs*).

'Normal' people (not being ecologists) want no disturbance: for example they want to have a chance to look at deer. Deer, however, in our present world need forest management (= disturbance). The forest requirements for a good population of deer (such as *Betula* and *Tsuga* stands) are produced by management (harvest and reafforestation). When, in such a case, politicians or policy-makers are asked for decisions, the danger is that expedient decisions cannot consider all aspects of the ecosystem. For example, social aspects should be included, as they are by Jacobs. (*Post*).

In this context it should be kept in mind that agricultural systems can be stable only with the support of the diversified structure of human society. Primary production is these systems is often compared by using the efficiency coefficient of solar energy conversion into net-primary production. Alberda, de Wit, Niciporovic, as well as other authors, have indicated that this coefficient lies between 5–10% Ph.A.R. (*Ph*otosynthetically *A*ctive *R*adiation) in intensively managed field cultures. If the total energy subsidy to these systems were taken into account along with solar energy, we should certainly arrive at appreciably lower values for the efficiency of the total percentage. (*Kvet*).

This is in full agreement with present opinion. It should be added that many ecosystems maintain a high level of photosynthesis with a low input of energy other than solar radiation. However, in the case of input of energy in the form of fertilizers, it is justifiable to distinguish between application aiming at a replacement of 'harvest losses' and application aiming at growth optimization. The latter type of treatment should be looked at more critically than the former, because an important part of the problem's solution may be recycling. (*Tamm*).

Some questions arise about man's management in creating water reservoirs (such as Lake Nasser and other man-made lakes). Such 'systems' cannot be converted into 'natural ecosystems' because of the continuous, unpredictable and unnatural fluctuation in water volume. Where no 'natural' rhythm occurs it is at least doubtful whether these reservoirs can ever reach a biologically controlled state of stability. (*Curry-Lindahl*).

Such conditions can, sometimes, lead to a rapid initial rise in fish production, but there is no guarantee that later on an invasion of parasites will not occur, including *Schistosoma* which causes the disease bilharzia in man. Lake Nasser (in Egypt) has already caused concern about its effect on the health of neighbouring human populations. (*Kohlemainen*).

It is evident that in such cases there is a need for temporary or continuous human intervention. In a multiple causal network with many independent parameters, the ecological 'rule' is, at best, the hyperspace frame of possibilities within the network. Each individual ecosystem may follow quite different paths within the frame. (*Jacobs*)

Under such conditions, and even more under so-called polluted conditions, as in the great lakes of Canada, a 'rule of thumb' might be that phosphorus (P) is of primary importance in regulating eutrophication. This is a general rule for all freshwater lakes, and probably for other water systems. (*Likens*).

The recent documentation of the role of phosphorus represents one area where ecologists can make recommendations based on hard data. Several studies have demonstrated that P-control is very effective in combating eutrophication (cf Lake Washington, Edmondson, Science, 1970; Little Otter Lake, Mechalski & Conroy, 1971; Shagawa-lake, Malueg, 1974; the general problem, Schindler, Science, 1974). (*Schindler*).

But how do we avoid the dangers of decision-makers adopting over-simplified rules-of-thumb and applying them to too wide a range of situations? (*Le Cren*).

One of our tasks is to emphasize that there are few general rules, but many rather specific ones, and to explain why this is so: that a network of numerous non-linear functional relationships exist in a system which has no 'brain' or centralized control. The more relationships between organisms are obscure or complicated (and they certainly are), the more quasi-interdependents may intervene. (*Jacobs*).

Man has also had an important impact by the construction of very resistant (chemically) artificial compounds (the plastics). How far are micro-organisms able to recycle rapidly any form of these. (*Scossiroli*)?

Some plastics and other polymers (but very few) are subject to biological degradation. For these few man-made compounds microbial destruction is possible, and hence biological degradation processes are certainly feasible. For all other (most) synthetic polymers no good evidence of microbial activity exists. Hence disposal by biological means seems most unlikely. (*Alexander*).

Human perturbations have unknown results. In this context it is evident that domestication of plants (for example in agriculture in tropical regions) is sometimes followed by an increase in plant pathogens. These observations, corroborated by the problems arising from the establishment of ranches for cattle breeding in tropical Africa, call for studies on the effect of human intervention. (*Mukiibi*).

These questions are not easily answered. On the one hand pathogens are a group so specialized that no general system can yet be formulated. (*Alexander*).

On the other hand it should be remarked that the introduction of cattle breeding in the tropics without any study of how to use the proteins from the natural system, is no real solution. It is evident that cattle (not even the best breeds) do not grow in such a way that the human population could not get more protein by using the indigenous fauna. (*Vlijm*).

What can be said about the projections regarding the effect of the anticipated 'increased heat' upon the activities of microorganisms, globally, more specifically considering these types of disturbances over the next 25 years? Physiological processes of aquatic organisms are greatly influenced by temperature. Increased temperatures considerably below lethal levels can modify growth rates, reproduction, distribution and thus modify productivity. The extent of these changes on a global scale are not documented by the principles in general. (*Hasler*).

The available data are confusing. Increasing the temperature as a rule increases microbial activity up to a certain point, but a rise in temperature may also alter nutrient availability, change the abundance of decomposers, predators and parasites, and affect oxygen solubility and the release of toxins. Hence it would be foolhardy at this time to make any projections on the response of those microorganisms concerned with biogeochemical cycles to the higher temperatures. (*Alexander*).

In general it can be asked how far models, as put forward by Harte, May and others, can be used in evaluation of systems. When the whaling industry is finished, can the surplus production of krill (*Euphausia superba*), which now amounts to 50–100 million tons/year, have any effect on other predators (fish, seals or birds)? (*Dunbar*).

The type of modelling shown here might have an application to these problems. The application of models seems to be extremely important and may provide indicators of stability and information about thresholds of instability hitherto overlooked. (*Harte*).

This leads to the question if it is possible to use this type of work and the eventual conclusions for helping policy-makers in coming to their policy. (*Evenari*).

These types of models and methods can be understood by scientific advisors and translated to policy-makers. This seems to be the appropiate 'information chain' in which scientific insights should flow. However, this does not mean that these types of models are self-explicable, either to lay-men or to politicians. (*Harte*).

Session 5

Strategies for management of natural and man-made ecosystems

Chairman: W. B. Banage

Strategies for the management of natural and man-made ecosystems

J. D. Ovington

Introduction

As community interest in the environment has grown there has been a progressive shift in the emphasis and pattern of resource use towards a more ecologically sensitive economy. No longer is economic merit seen as the sole criterion for the assessment of development projects but both social and ecological factors have also to be taken into account.

Matching natural resource use with environmental protection presents a formidable task made more difficult by the urgency generated by exponential population growth and the need to improve the living standards of the many seriously disadvantaged people who are already born. Answers are needed to a bewildering variety of problems occurring at all organisational levels—global, regional and local.

Ecologists with their interdisciplinary approach have much to contribute to the solution of these problems and to ensuring that ecosystems are managed effectively for greatest benefit. Ultimately any biological solutions have to be related to social and economic considerations and reconcile differences between sectional interests in the population.

The role of ecologists

In order that ecologists can evaluate or suggest alternative strategies of ecosystem management for different purposes they need to develop conceptual frameworks whereby to simulate ecosystems. In this way they can appreciate better how ecosystems function, the multiple ways the component parts and processes of ecosystems interact in different circumstances and the relationships between ecosystems constituting a landscape unit for planning purposes. Using such techniques, ecologists can foresee many of the implications of different management strategies without ecosystems being placed at risk.

As a group, ecologists also have a special responsibility for encouraging nature conservation as an acceptable form of land use. More than other scientists they must present the scientific case for the preservation of any representative ecosystems they regard as being valuable ecologically, whether natural or man-made, and which are endangered through changing circumstances.

Increasing recognition of the contribution that ecologists can make to resource use has meant a greater emphasis on applied ecology. Ecologists are now commonly included in national and regional planning teams and are being employed by development agencies to advise on environment protection and wise resource use.

The problems of applied ecology are compounded by the dynamics of ecological situations, and the different time scales which are involved. For example, the replacement of a natural forest by a commercial plantation will have an immediate effect on the flora and fauna present but long term changes in soil properties resulting from intensive production may be of greater significance in relation to sustained production and the biological potential of the area. Even longer term changes may be taking place in the biosphere, the ecological implications of which are largely unknown. Typical of these is the evidence of change in the world climate with the apparent displacement of the belt of monsoon rainfall southwards. This has important ecological consequences and could affect the world pattern of biological productivity.

Ecological principles

Since ecology is a young science, ecological knowledge is regrettably incomplete with relatively few competent ecologists available to meet the challenge. In practice, many ecological judgements may not have the factual background desirable under ideal conditions of decision making. Consequently some ecologists have directed their energies to elucidating ecological principles as guidelines for development (Dasmann et al., 1973).

Ecological evaluation of past development schemes, whether the schemes are regarded as successful or disastrous, is important in refining ecological principles. In this way it is possible to ensure past mistakes are not repeated and the lessons of successful management learnt. Because of the inadequacy of ecological knowledge and the complexity and uncertainty in making sound ecological judgements, management strategies must be tentative and incorporate environmental monitoring and flexibility of management. As experience is gained, management strategies may need to be modified.

Fortunately the concept of ecosystem (Tansley, 1935) and the somewhat similar concept of biogeocoenose (Sukachev & Dylis, 1964) provide means of integration whereby to bring together different facets of ecology. They have also stimulated synecological investigations and encouraged interdisciplinary study.

Nature of ecosystems

Ecosystems may be regarded as basic management units for development purposes. In general, ecosystems contain a diversity of living organisms and these organisms display in their life processes a complex pattern of interdependence and interaction with one another and with their environment. Ecosystems are rarely static, their boundaries are subject to change and they may vary both seasonally and progressively with the passage of years. The factors determining the stability of ecosystems are not fully understood but it appears that increasing complexity enhances resistance to rapid change arising from natural causes.

Virtually all ecosystems have been modified by man, either deliberately or accidentally. Whilst some terrestrial ecosystems, particularly in remote places, can still be regarded as little affected by man, others such as commercial forest plantations, agricultural fields, and cities are highly artificial and usually less diverse biologically than the preceding more natural ecosystems.

With his large numbers and advanced technology, modern man has acquired an immense capacity for massive and rapid modification of ecosystems. New landscapes are being created constantly. Where such exploitation leads to widespread vegetation destruction followed by soil loss, there is serious disruption of economic processes and loss of biological capital and production potential.

However development and change are not inherently bad. In fact man has been remarkably successful in developing management systems which have created artificial ecosystems capable of producing the massive amounts of goods and the variety of services demanded by modern civilisations. Many of these artificial ecosystems are relatively stable provided certain inputs, e.g. of fertilizers, can be maintained. Some artificial ecosystems are regarded as being of high conservation value by virtue of the numbers and variety of organisms they support and because of their cultural and aesthetic interest.

Management techniques

Ecosystem management may involve intervention in three main ways which are detailed later, but all bring about changes in ecological succession, biological productivity and in the nature and magnitude of the four dynamic processes whereby ecosystems function as integrated biological units. These processes are organic production and break-down, energy flow, nutrient circulation and water movement.

The most common method of ecosystem management by man is to change the species of living organisms present. In extreme cases, as in intensive agriculture and forestry, this involves the complete replacement of the natural flora and fauna with exotic species selected and bred for their capacity to produce materials valuable to man. Commonly this results in increasing simplification of ecosystem structure and uniformity of vegetation cover to facilitate harvesting of the crop plants or animals. In a less extreme form the natural species may be retained but their proportions are altered, for example, by selective harvesting.

The second method is to modify ecosystem structure, usually in order to concentrate productivity in a particular plant layer. In the case of grazed pastures, photosynthesis is restricted to the ground-layer plants which are available to grazing stock. In contrast, management in intensive forestry usually aims to attain maximum wood production by having the relatively high tree canopy as the main photosynthetic layer and with a minimum of herbivorous grazing.

Decomposition takes place at ground level. The forester may regulate the configuration of the tree stock by thinning to direct the photosynthate into particular stems and by selective harvesting to produce an uneven forest cover with a range of tree size classes.

The third management technique is to change the ecological conditions. A whole array of new methods has become available through modern technology. These include the mass use of chemicals as fertilisers, insecticides and herbicides. Fire is an old but nevertheless effective means of changing ecological conditions, and its use has been refined with the introduction of aerial fire bombing. Climatic modification e.g. by shelterbelts; manipulation of the water resource by drainage, irrigation or flooding; and soil treatments such as ploughing because of mechanisation can now be done on a scale that would have been impossible a century ago. Mass modification of topography with developments in earth moving equipment has become a practical proposition and radical transformation of land forms now occurs, associated for example, with road building and airport construction.

Much environmental change arises inadvertently, e.g. the removal of the tree cover in Western Australia unexpectedly led to increased salinity of ground and stream water. The burning of fossil fuels is indirectly increasing the carbon dioxide content of the air and possibly radically affecting the world climate. Similarly the construction of tourist roads around Ayers Rock in Central Australia proved disastrous ecologically in that the roads blocked the natural water drainage pattern causing extensive death of the vegetation cover. Ironically, these tourist roads diminished the attraction of the area for tourists.

Public concern for management effects

Inevitably Man's use or manipulation of ecosystems results in change, and any management of ecosystems must take into account public reaction. Whilst populations were small the modification of wilderness areas and their conversion to profitable agriculture or forestry were seen as beneficial. Now developers are assailed by environmental groups who base their criticism on the adverse consequences of past development schemes oriented to short term gain and without a long-run perspective. The public in general is becoming increasingly concerned to ensure that decisions about resource use are not restricted to a few people with vested sectional interests but are exposed to public comment and scrutiny. Broad-based community control is seen as being most effective in deciding whether on balance a particular land use development is beneficial and devoid of any unacceptable environmental consequences such as serious deterioration of the quality of air, soil, water and scenery.

Whilst it is generally recognised that ecosystem management is necessary to support the multiplying and more demanding population of the world, there is concern to ensure that the capacity of the earth to provide aesthetic satisfaction and sustain biological production is not irrevocably reduced to the detriment of future generations. Probably the scale and incidence of man-induced change is accelerating and the likelihood of an ecocatastrophe is increasing as human influence becomes more diverse and all pervading.

There is also pressure to set aside from development natural ecosystems to hedge against future uncertainties. The destruction of natural ecosystems and species extermination are increasingly seen as inherently wrong on cultural, religious, aesthetic and scientific grounds and as an unjustifiable erosion of the human environment. In many cases the finance devoted to attempting to change critical areas of poor fertility to highly productive management systems would have been better invested in increasing the yield from existing intensively managed ecosystems.

Changing public attitudes and increasing awareness of the issues involved are important factors to be taken into account in planning a strategy of ecosystem use. Schemes of ecosystem use are less likely to be applied if they fail to gain public support. Because of growing public concern with environmental matters, entrepreneurs are being increasingly constrained in development projects and government and international agencies are being required to take a broader, more integrated long term approach in allocating land as between alternative forms of use. Massive educational programmes directed to influencing public attitudes to resource use have been mounted by a wide range of interested organisations.

Decision criteria for ecosystem management policies

It is unrealistic to expect that there will not be continuing transformation of ecosystems. These transformations will occur in a variety of ways, some will be insidious but all pervading as happened with the spread of toxic chemicals with the development of new insecticides. Other transformations will be more obvious but local, as with the use of defoliants in Vietnam.

Clearly with an increasing population and mounting social pressures to improve living standards, further conversion of many relatively natural ecosystems to highly artificial ecosystems can be anticipated. Equally, the management of existing agricultural areas may be intensified as has happened with the 'Green Revolution' or as major development schemes, such as the Mekong River Development Plan, are completed. Less reliance may be placed on secondary producers as plant protein replaces animal protein for human consumption.

Some ecosystems may undergo a rapid transformation with exploitation followed by rehabilitation. For example, sand mining companies are now being charged with the responsibility of restoring a dune topography after working over and levelling coastal sand dune areas for the extraction of valuable minerals such as rutile and zircon. The dunes are artificially revegetated with ecologically appropriate dune species.

Despite pollution control measures, some industrial pollution of the environment is inevitable. The waste engendered by the growth of urban conglomeration poses immense disposal problems.

The fundamental question is whether, at the one extreme, these human impacts are likely to be of such a nature and scope that they are unacceptable because they are so detrimental that they place the survival of the human race in jeopardy. Alternatively it is possible that the impact of man can be accommodated within the foreseeable future until the disturbing influence of man can be brought into a more stable and intimate balance with the global environment realities.

This relationship between ecology and social action as a basic criterion of decision making in ecosystem management is most important and has been elaborated by Commoner (1973). He sees the basic choices as 'Blind application of ecological principles to human society and the making of the effort well enough so that we can devise new ways of fulfilling them which are consistent with human purposes or slavish acceptance, in the name of ecology, of a rigidly controlled society, and the freedom to choose, on the basis of both ecology and humanism, how we would live on this earth—between ecology and social inaction, and ecology and social action'.

Significance of planning for ecosystem management

As civilisation has evolved, Man's manipulation of ecosystems has been a changing process which has grown in capacity and variety in a disorganised way as the nature of decision makers has changed. Some people would suggest the price of advancing civilisation in such a manner has been too costly and wasteful, the benefits gained by one generation being at the expense of later generations. Other people would argue that it is past technical advancement and exploitation of natural ecosystems which permit man to be able to indulge in activities other than those of providing food and materials e.g. in setting aside from further development certain ecosystem types.

However, the past is forever, and unchangeable. The essential need now is for purposeful planning as a basis for the orderly and efficient management of ecosystems and for the improvement of environmental quality. Whilst the problem is ultimately global in character, it is unrealistic to envisage the immediate establishment of an effective supranational organisation controlling development on a world basis.

The difficulties encountered in achieving international agreement for example on the conservation and management of whales are indicative of the problems likely to arise. Global ecosystem management is even more complicated and difficult to achieve than rational management of the whale population because of the marked disparities between different regions in biological capabilities and human affluence.

Currently, tropical forest ecosystems are being exploited and destroyed wastefully to provide agricultural land, to meet the timber needs of industrialised countries of the temperate region and to provide finance to stimulate modernisation of tropical nations. Ideally it might seem desirable to regulate and coordinate the management of tropical forests after identifying their most appropriate role in global environmental dynamics and trade, but such planning would be meaningless unless the aspirations of the inhabitants of the tropical zone are fully protected.

Whilst the task of developing large scale schemes of resource use is formidable, it must not be forgotten that new and powerful tools have become available e.g. photography from aeroplanes and spacecraft and the growing capacity for handling masses of data provided by the development of computers. Furthermore, there is now considerable potential to inform the public and to stimulate communal action because of the information potential provided by television, radio and other mass media information services.

However, to be effective the overall strategy of ecosystem management needs to arise from detailed local planning studies e.g. of water catchments from which some integrated regional and global strategy might emerge. These local studies should be carried out by multidisciplinary teams and would be concerned primarily with the assessment of the suitability or capability of ecosystems to meet various needs under different management regimes and for different purposes e.g. agriculture, forestry, recreation and conservation. Particular attention would need to be paid to the possibilities of multiple use and rotation of use e.g. for some ecosystems it might be appropriate to have long term rotation of forestry and agriculture in order to maintain soil fertility.

Four distinct activities would be involved in carrying out these surveys:

1. the collection of basic data relevant to potential uses of the ecosystem;

2. analysing the collected information (ecological, economic and social) particularly in terms of capability to fulfil different uses under different managements;

3. presentation of alternative management regimes with their advantages and disadvantages to the public so that people can express their preferences;

4. implementation of the agreed management with suitable monitoring and provision for periodic review of management to accommodate to changing circumstances.

The complexity of ecosystems is such that in practice the surveys would necessitate common sense compromises between comprehensiveness, accuracy and the limitations of time and staff. Attention would also need to be directed to the availability of people having the necessary management skills, for it would be pointless for instance where management skills are scarce to prescribe complicated management regimes which could not be put into operation.

Conclusion

Attention has been drawn to the significance of some changes that are taking place in the ecosystems of the world and of the need for proper care and imagination in developing comprehensive strategies to ensure that their management is soundly based and directed to meeting acceptable objectives. A need is seen for an hierarchical system with ecosystems being the basic units from which to develop a comprehensive strategy for the management of natural and man-made ecosystems. In this way more rational land use decisions with minimum ecological risk can be made and local and regional viewpoints can be reconciled into overall schemes of natural resource use and development.

Ecosystem management cannot be based solely on ecological considerations but social and economic factors must also be taken into account. Because of mounting public concern with the environment, the implementation of any management must be based on public scrutiny and support.

References

Commoner, B. 1973. Ecology and Social Action. 13. The Horace M. Albright Conservation Lectureship School of Forestry and Conservation. University of California, 27 pp.

Dasmann, R. F., Milton, J. P. and Freeman, P. H. 1973. Ecological Principles for Economic Development. John Wiley, 252 pp.

Sukachev, V. and Dylis, N. 1964. Fundamentals of Forest Biogeocoenology. Oliver & Boyd, 672 pp.

Tansley, A. G. 1935. The use and abuse of vegetational concepts and forms. Ecology, 16: 284–307.

Author's address:

J. D. Ovington
Department of the Environment and Conservation
Canberra
A.C.T. 2601
Australia.

Notes towards a science of ecological management

C. S. Holling and William C. Clark

The thesis presented here is quite simply that it is now possible to catalyse a new science of ecological management/engineering. The need is obvious, but most significantly the essential pieces, independently developed, can now be integrated and/or used on ecological problems. Even more important, a relatively new concept emerging from ecology can provide a conceptual focus for a new regional strategy of ecological and resource management.

Now that the more intemperate extravaganzas of the recent concern for ecological issues have passed, it becomes possible to identify some solid foundations for ecological management science. On the ecological side, these lie in three areas which have been developed over the past fifty years. The first two have come from applied areas—insect pest ecology and fisheries ecology. Both have been characterized within a rich scientific tradition, one which comes as a surprise to those more familiar with the 'eco-freak' image of recent years. Here, there is a remarkably sound empirical base—both extensive and intensive—characterized and indeed initiated by the R. A. Fisher school of statistics and sampling theory. There is also a mixture of laboratory and field experimentation which has unravelled and generalized many of the key causal relations which link organisms with each other and with their environment. And, finally, there has been an active mathematical tradition of modelling; differential equations initially, and then—with the appearance of computers—differential-difference equation mixes leading up to but as yet not beyond simulation models.

Simulation models in ecology, as in many fields, initially were oversold. There were noble and grand efforts to develop *the* generalized model of this, that, or the other ecosystem. Many models became so complex as to be as mysterious as the real world. We are now through that inevitable stage and we see growing numbers of effective efforts to bound, intelligently, problems from the outset, to compress and simplify up to but not beyond the point where essential behavior in space and time is retained, and increasing effort to interrelate the modelling with rigorous field and laboratory experimentation.

The third area of relevant ecological development is the theoretical. Ecological theory has tended to be divorced historically from application and finds its roots more in evolutionary biology. But from that theory have emerged a number of concepts of ecosystem structure which have begun to form a happy partnership with the empirical and modelling approaches of the applied branches of ecology. The result has been several major steps towards describing and quantifying the stability behavior of perturbed ecological systems. It is this latter development that potentially provides the conceptual foundation which gives us the temerity to suggest that something new and innovative is possible in designing a science of ecological management. We shall amplify this point later.

Now, however, it is more important to touch on the missing pieces of this apparently glowing story. And the missing pieces represent the serious gaps which have made ecologists lousy managers. The main issue is that man and society have largely been left out of even the best of applied areas. It is true that economics (in its guise of resource economics) has crept into fisheries management. The partnership flowered for a time but began to wither as the economics tended to move into more and more esoteric academic numerology. There are notable exceptions, but the fact remains that the marriage of ecology and economics has been an uneasy one which has, with few exceptions, never been effectively consummated. The reason, we believe, is that the marriage was largely in isolation from the broader concerns of society and from the techniques which have evolved in the management sciences, particularly policy analysis and decision theory. The result is that applied ecology has tended to be descriptive and not prescriptive. Hence the new conceptual focus should illuminate an integration of the best of ecology/economics modelling, policy analysis, and decision theory to provide the basis for a new science of ecological management/engineering.

Let us now touch further on the relevance and need for a fresh conceptual framework. The past management of ecosystems has implicitly presumed that the consequences of an incremental action will be quickly detected. If the intervention produces higher costs than benefits, then a revised incremental action can be designed. It is this trial-and-error strategy which has succeeded in producing phenomenal increases in production of the food, fiber and other resources needed by man. Little knowledge of ecosystems was required so long as the consequences of an erroneous trial were minor and alternate trials remained possible. It has been an admirable and effective method of improving our lot in spite of our ignorance.

But now incremental acts seem to be producing more extensive and intensive consequences, consequences which resist further incremental solutions. The geographical scale of our interventions and their magnitude can now make an erroneous trial disastrous. That is dramatically obvious in nuclear power developments, but it is equally true of resource developments. In addition, other consequences are emerging from the accumulation of past incremental decisions. Our remedial responses to these new emergencies are as shortsightedly ad hoc as their original causes. Banning DDT may seem admirable, but advocating such narrow solutions can lead the ecologist to join that group of apparent villains (we emphasize apparent) who planned our freeways and designed our dams. That is a good way to destroy the myth of the ecologist's moral rectitude but hardly a way to be responsive to significant social needs.

Trial-and-error seems to be an increasingly dangerous strategy for coping with ignorance. And yet the solution cannot be to withhold action until we have sufficient knowledge. We need a new strategy for dealing with the unknown. One direction to go might simply be to engineer nature, (i.e. the unknown) out of the equation. With enough concrete and energy we could make the world a known one. That is the route which led to the semi-humorous suggestion that the pest problem of 'miracle' rice could be resolved by paving and then flooding all of southeast Asia. But we don't have enough concrete and energy, and there is no way to engineer out those vexing and disturbing human demands for 'quality of life'. That scarcely is the route for dealing with unknowns.

Four major classes of uncertainties and unknowns may be identified. We have incomplete, although growing, knowledge of the functional relationships within ecosystems—of their number, kind, form and intensity. Also, we have limited knowledge of the social objectives for ecosystem management. There are hidden objectives and they remain so until they are suddenly no longer satisfied. These two sources of ignorance—the descriptive and prescriptive—are important but manageable. Present techniques can identify and hedge against these sources of uncertainty in inputs, parameters, functions, and alternate values. Much of systems analysis is directed to these problems.

But what of the qualitative unknowns inevitably dealt us by 'fickle fortune'? The basic rules underlying linked economic-ecological systems can change. Unexpected species can suddenly appear and dramatically alter ecosystem structure. Unexpected economic changes can do the same—witness the observed and potential impact of the energy shortage on food production. And the one-in-a-thousand-year flood or drought is as likely to occur this year as any other. In the same way, prescriptive aspects of management can experience equally unpredictable changes. Human objectives which seem so clear at the moment can and do shift dramatically, leaving society committed to policies and systems which cannot themselves shift to meet these new needs.

Few systems which have persisted for extensive periods exist in a state of delicate balance, poised precariously in some equilibrium state. The ones which are, do not last, for all systems experience unexpected traumas and shocks over their period of existence. The ones which survive are explicitly those which have been able to absorb these stresses. They exhibit an internal resilience. Resilience, in this sense, determines how much disturbance—of kind, rate and intensity—a system can absorb before it shifts into a fundamentally different behavior.

Historically, ad hoc management approaches have succeeded specifically where applied to highly resilient systems. The inevitable mistakes, made from ignorance, were just additional disturbances which could be absorbed by the resilience of the system. But that resilience is not infinite. We can now show, from our ecological models and from real-world examples, that ecological systems are multi-equilibria ones and, moreover, can demonstrate the causal mechanisms leading to multiple equilibria. These equilibria are bounded and so produce stability regions within which the variables fluctuate and move with relatively weak damping. Exogenous disturbances—natural or man-made—generally cause modest or undetectable numerical change within this highly fluctuating world. The qualitative behavior remains unchanged and, most significantly, no signal is generated of a possible contraction of a set of commonly inhabited stability regions. That signal is only generated when the disturbance is great enough to flip the system into regions not normally occupied. Or it is generated by accumulation of past incremental decisions which have led to a contraction of the normal stability regions. A disturbance, such as a normal fluctuation of climate, which previously could be absorbed can now no longer be absorbed. That is what much of the eutrophication literature is all about; and that is what has led to the collapse of most of the freshwater fisheries of the temperate world. A more detailed treatment, with examples, can be found in Holling (1973).

The point we wish to make is that the traditional view of stability, as presently practiced, concerns responses to small perturbations and considers as stable systems those which fluctuate least and damp most rapidly. But an equally valid view concentrates on the responses to large perturbations and reveals that highly fluctuating systems can be immensely 'stable' in that they can persist in the face of major disturbance.

This view leads to a strategy of management that can attempt to work with the natural dynamic rhythm of ecosystems, that attempts not to eliminate fluctuations but to transfer them into directions less in conflict with man's desires; that attempts to design systems which are not so much fail-safe but safe in the inevitable event of their failure (remember Hurricane Agnes?).

With that rhetoric behind us, let us attempt to encapsulate the ingredients of this new science of ecological engineering.

1. Conceptual: a rigorous development of the resilience/stability concepts based on representative theoretical and applied models, ranging from coupled differential equations (for historical reasons), through simulation models of simple ecological systems (few state variables), to those of complex ecosystems (many state variables, non-linear, spatial disaggregation).

2. Numerical quantification of resilience: the ecological 'Reynolds' number(s); retrospective case studies from ecology, resource sciences and social sciences analyzing the resilient behavior of the systems in response to major stress.

3. Development of resilience indicators which provide at least surrogate measurements reflecting the size and nature of stability regions. Such indicators seem to fall into three main classes: resilience in unused environmental 'capital', resilience in relation to stability boundaries, and resilience to policy failure.

4. Development of environmental standards which recognize the fluctuating nature of systems and lead to a balance between preventative and remedial responses to meeting standards (see Fiering & Holling, 1974).

5. Development of a strategy for generating policy alternates ranging from the 'fail-safe' to the 'safe-fail'.

6. Blending the above with existing and expanded techniques of systems analysis which have been so effectively developed in the water resource field in particular: in essence all those techniques of policy analysis including optimization (where it can be stretched) and more heuristic 'dirty' techniques.

7. Joining the above, in turn, with decision theory to deal with questions of decision making in the face of uncertainty, with techniques of multi-attribute decision making, and with techniques which assess the way individuals evaluate quality of their environment.

8. Finally, developing communication formats and processes which force the analysis to be responsive, useable and transferable to the man who makes decisions and those who endure those decisions.

All this, we hasten to add, should be developed around carefully chosen case studies and field experiments which possess both applied significance and the potential for conceptual and methodological advances.

References

Holling, C. S. 1973. Resilience and stability in ecological systems. Ann. Rev. Ecol. Syst. 4: 1–23.

Fiering, M. B. and C. S. Holling, 1974. Management and Standards for Perturbed Ecosystems. Research Report R-74-3. International Institute for Applied Systems Analysis, Laxenburg, Austria.

Authors' addresses:

C. S. Holling
Institute of Resource Ecology
University of British Columbia
Vancouver
Canada

William C. Clark
Institute of Resource Ecology
University of British Columbia
Vancouver
Canada

Management strategies in some problematic tropical fisheries

W. H. L. Allsopp

Introduction

There is increasing public awareness of the importance of the environment in the tropical world. Ecology is no longer an ill-understood technical term used by a few fishery administrators or policy-makers in tropical countries. Thirty years ago tropical fisheries development was almost synonymous with indiscriminate exploitation of unknown resources. The period of 'management' involving so-called logical controls is gradually changing to systems which introduce corrective measures justified on the basis of assessments. Speculative management is progressively giving rise to policy decisions based on some scientific data; rationalisation is replacing adventurism while responsible proprietorship is now seeking ecological data that can provide the tools for effective management measures.

However, the multispecific nature of tropical fisheries resources complicates the management problem on biological grounds. Furthermore, because of the demand for fish as an essential food, rather than as a luxury item or high-priced delicacy, in the developing countries of the tropical world development imperatives impose social and economic obligations which require immediate decisions. The inadequacy of data, of trained technicians and of the general means to-accomplish effective management, compound the problem in a relentless time-frame of urgency.

The problems that confront tropical fisheries have generally resulted in there being no clear management strategies. The complexity of the circumstances and the imperatives of providing food for subsistence or even survival of peoples in many circumstances makes management, in the true sense, impossible. There are few closed systems, though aquaculture is gaining increasing importance. There is need for the establishment of practical strategies if management is to be achieved.

This paper is intended to illustrate some of the complex problems that confront tropical fisheries management in conjunction with many priority obligations of administrations who have the responsibility of promoting human well-being in developing countries while causing the minimum of environmental degradation. The examples chosen are considered to be of significance to the international community of science.

The role of tropical fisheries

The slogan of the sixties of 'Freedom from Hunger' resulted, among other things, in deliberate efforts to improve food resources in all countries of the world. Fisheries were therefore required to produce more food, particularly in the emergent nations, and the world's production of fish increased from 19.6 million tons in 1948 to 65.6 million tons in 1972. Of this, production from tropical and sub-tropical waters represented approximately 20% of the global catch in 1948, compared with 36.8% in 1972 (FAO, 1972).

With human food as the objective, and maximum sustainable yield as the precept, the rationale has been changing to the maximum economic yield, while development action has taken place in many cases even before all the facts were available to justify commercial investments. The recourse to systems of speculative management resulted as developments were outpacing the availability of scientifically adequate data translatable into management measures. Where intensive fishing was taking place, some investigations gave clues to the need for diagnostic scientific monitoring, biological data collection and the need for an early warning system. In situtations where the need for management was recognised as urgent, the problems of management were completely without precedent in scale and complexity for those who were charged with that responsibility (FAO, 1970).

The dilemma of choice between the need for maximum food and the desire for a rational level of harvesting was often resolved by a decision to invest and expand after initial entrepreneurial success gave some promise of satisfying the food, employment or other short-term economic and social needs. When there were evident indications of harmful ecological consequences, some corrective measures were considered. In general the available data were limited. If these indicated that limitation of production was necessary, they would be challenged before they could be transformed into applicable measures and these measures might not be implemented even if clearly indicated as necessary precautions. The initial decision to opt for food, with the burgeoning populations and employment problems in the developing countries, becomes an almost permanent and hard-to-reverse decision when the immediate benefits are weighed against the long-term and possibly permanent ecological difficulties.

While fishery ecologists in the developed world have for decades tried to convince the public, the impact of their advice is only now becoming acceptable. In the tropical world with few, less potent voices, and with only limited data to support conservation advocacies, problematic tropical fisheries have been facing serious and little appreciated problems that tend to create an increasing ecological imbalance through varied circumstances.

It is therefore evident that the mere indication of probable difficulties is not a deterrent to possible mal-practice in a fishery of a developing country. Some effective economic alternative which provides a satisfactory interim solution will be the only acceptable procedure. A strategy for ecological conservation should therefore involve clear evidence of some or all of the following advantages: public health improvements including control of waterborne diseases; employment; diversified food supplies; economic benefits such as greater total yields; satisfactory profits.

Measures which seem to be more in the interest of the fish, with no immediate benefits for man, though of recognised importance, are not convincing enough unless the urgent needs of man's food supply are first satisfied. In the face of local scientific advice there are also exploitive external pressures, and it is sometimes found that while restrictive measures are applied within the national fishery, no measures are being applied to foreign high-seas fishing, which crop the same stocks and resources with impunity (FAO, 1971).

Tropical freshwaters

Flood-plain fisheries. Africa's inland fisheries provided in 1972 about 1.20 million tons of fish, mainly from the large river basins and lakes. The main source of fish supplies of these river basins are the flood-plain fisheries such as those of the Niger in Mali, the Zaïre, Nile and Zambezi rivers. Fishery regulations dating before the sixties controlled mesh sizes and in some cases established fishing seasons which are now largely ignored. These regulations were logical but somewhat speculative, and in most cases not based on adequate data. Decline in the catch cannot be attributed only to increased fishing pressure, though this is clearly contributory. The great losses of fish stocks occur in the seasons when the isolated pools, ox-bow lakes and marshes lose their fish through drought and predation. Comprehensive surveys are needed to determine the size of species to be harvested, which fisheries areas may best be cropped, and to what extent other localities may be conserved. The vast inland 'delta' of the Niger supplied 100 000 tons of fish annually. The Sahelian drought has since compounded the problem through exceptionally low water levels; fishing pressure and natural predation have reduced the parent stock to unprecedented low levels in the Niger, Chad and Nile basins.

There are hundreds of fish species with varied ecological niches, habits and different growth rates. There is variable fishing pressure for subsistence fisheries and semi-industrial activities. Distances to markets or fishable areas, seasonality of supplies of types with higher consumer preference, and fishing gear availability, introduce many variable factors that make an inflexible management policy impractical and unacceptable. How is the dilemma of management of flood-plain fisheries to be solved? There is clear need for the development of an effective strategy based on practical and proven experimental systems.

Such tropical flood-plains are also a feature of the Indian sub-continent, S.E. Asia and Latin America. The vast permanent flood-plains of the Amazon are similar to those of the Zaïre River, while smaller flood-plains in the Essequibo River in Guyana also typify these problems. Since the flood-plain fisheries represent one of the greatest under-utilised tropical fish resources, ecological research should be concentrated on these (Allsopp, 1974).

Lake fisheries. Eight major African lakes have been studied intensively by teams of scientists supported by U.N. Development Programme Funds (Obeng, 1969). Of these, five were man-made lakes of very recent history, created essentially for hydro-electric power and irrigation supplies. These are among the first lakes created artificially in the tropics, and as such a close study and monitoring of their ecological evolutionary processes are of considerable world-wide scientific significance. However, their fish resources have been a matter of considerably less significance.

It is here suggested that the recognition of the importance of ecological problems of tropical man-made lakes has resulted mainly from the impact of problems created by aquatic weeds. These impeded hydro-power generation, transport, affected resettlement habitations, sheltered water-borne disease vectors and accordingly increased development costs. The control of aquatic weeds thus made fisheries of these lakes almost as important a consideration as the hydro-power, irrigation or transport purposes which the lakes provide. Though it may have been a secondary consideration, the fishery potential of these lakes has assumed increasing significance to policy makers and it is now generally appreciated that fish and other organisms contribute both through ecological equilibrium with effective management intervention, and as a continuing food source. The role of fisheries management is nevertheless only a part of the multi-disciplinary activities involved in the effective development of these artificially created tropical lakes.

The management of the ecosystem of these tropical lakes involves control measures of human disease associated with aquatic vector organisms, (bilharzia, onchocerciasis, malaria, filaria, etc.) which are linked with the explosive growth of aquatic plants that provide shelter for the breeding of intermediate hosts. Apart from this, limnological studies, productivity and eutrophication provide unique and varied problems. A great variety of fish species is involved, while initial eutrophic conditions caused by the rotting vegetation and available nutrients support high populations of forage fish species before the population settles to natural equilibrium. Fishing is difficult because the tropical hardwood trees, of the submerged forests which do not decay under water, provide sanctuary for fish that cannot be gill-netted, trawled or easily trapped. Ecological studies must continuingly involve a close monitoring of biological data. Again, costs, personnel and other priorities create an administrative dilemma, resulting in developments outpacing management measures or capabilities. These African lakes are vast closed systems of greater complexity for management than are the hydro-electric impoundments of temperate climates. The development of effective management systems is a considerable scientific challenge which should be met by world science.

Control of aquatic vegetation. In other water bodies, whether lakes or rivers, agricultural developments involving the use of fertilisers, land clearance etc. have resulted in progressive but evidently permanent aquatic vegetation blooms, both of submersed and emergent types. The control of these in tropical countries was first achieved by manual labour, mechanical devices and subsequently by chemicals. The costliness of the first two made it impractical, the consequences of the last in conditions where the water supply is used for domestic purposes, make their use no longer acceptable (Manatee Workshop, 1974). The permanent stands of aquatic plants cause higher human disease incidence and lower production of fish. The prospects of natural conversion of the weeds, either by herbivorous fish, manatees (*Trichechus sp*), or terrestrial animals seems to be the only viable, though partial, solution. Much remains to be done to perfect systems for converting such world-wide nuisances as *Eichhornia* and aquatic grasses into animal food or economic products. Everything that contributes to a solution to this problem is desirable but if the weeds can be safely converted into comestible products (whether manatees, ruminants or fish), the solution is more likely to be adopted quickly and extensively. Chemicals are now too costly both to buy or apply, the areas are too vast, and the residual effects on the minor organisms of the food chains can have lethal and sublethal effects on fish, and probably cumulative effects on man. Wherever the water is used by rural peoples for domestic purposes, chemical control of aquatic plants is quite unacceptable.

It is therefore submitted that research and ecological studies of tropical floodplain fisheries, man-made lakes and tropical aquatic plants should be given priority attention. The proposal for the establishment of an international centre for research on manatees in Guyana deserves strong support, since the reproduction of large numbers of manatees can contribute substantially to the abatement of tropical aquatic weed nuisance, while providing more edible animal protein. Should the dominant climax form of these tropical waters be nuisance weeds or useful animals?

Fish culture. Fish culture represents one of man's oldest efforts at ecological manipulation and environmental management of fish. This husbandry practice has initially been of wild species with the progressive domestication of certain species (FAO, 1974). As a result of centuries of interest there are still only three major species-groupings of fish that are now virtually domesticated. These are (a) the carps (European, Chinese and Indian species), (b) the salmonids (salmons and trouts) and most recently (c) the catfishes (American and Asian).

The eventual aim of domesticating desirable food fish through aquaculture would achieve many direct and ancillary benefits including food supplies, environmental improvements, pollution controls and efficient use of national waters for maximum benefits. If, over the centuries, only three species groups have been semidomesticated, it may be wildly ambitious, even idle, to advocate fish culture research on all other suitable species. However, the comparatively rapid results obtained by the catfish entrepreneurs in the U.S.A. give promise that where the incentive and scientific endeavour is stimulated by profitable investments, the research results can be quick and lead to prompt industrial promotion.

There are hundreds of tropical fish species, the culture of which have been attempted in the tropical areas of the world. Rapid growth, desirable food habits as vegetarians or omnivores, adaptability to confinement, taste and market value, have been the selection criteria, among others. The juvenile stages were collected from the wild and in some cases progenitors were successfully reproduced. Aquaculture is clearly recognised as being critically significant for food supplies and the environmental improvement which supervenes. This is the most effective closed ecosystem for intensifying food production and yet ensuring environmental stability.

The major constraints to its extensive applications worldwide are (a) the supplies of juveniles of the important species, (b) the precise knowledge and availability of natural foods and artificial feeds required for effective growth and reproduction, (c) the effective control of diseases and parasites, (d) the genetics, selective breeding and hybridization towards total domestication of fish with the desired traits. Subsequently will follow the need to intensify the culture systems, to design installations, equipment and structural devices which are best suited to the species cultivated, while establishing industrialised aquaculture as a food production system of similar status to the commercial year-round production of poultry and pigs, in warm tropical climates (FAO, 1974b).

Research endeavours therefore need to be concentrated on reproduction physiology and larval rearing methods, together with nutrition requirements and the production of foods and feeds. The objectives should be to ensure adequate and reliable 'seed' supplies, independently of natural sources so as to lengthen the period of availability of juvenile fish for culture, and also to develop economic and nutritionally efficient feeds using locally available materials. By domestication of the species their genetic selection and hybridization will be facilitated. With large quantities of seed or juvenile fish, extensive under-utilised water bodies can come under active production, with conscious conservation practices.

Aquaculture improvements therefore, whether in freshwaters, estuaries or at sea, are an important strategic tool for maintaining environmental stability and the application of sound principles of intensive closed ecosystem management. The international coordination of aquaculture research efforts could help in a strategy of ecological conservation with a food production orientation (FAO, 1974a).

Tropical seas

East-Central Atlantic fisheries. Off the coast of West Africa in the East-Central Atlantic (from Gibraltar to the mouth of the Zaïre River), an important multi-species fishery has developed. During 14 years declared landings have increased 5 times (512%) and in 1972 stand at 2.92 million tons with fleets from about 42 countries participating. The fishing zones may be classified as northern subtropical and southern tropical with dermersal and pelagic stocks under varying degrees of exploitation intensity. The FAO committee for East Central Atlantic Fisheries (CECAF) has established a resource evaluation working group and endeavoured to introduce measures for improving and monitoring the stocks (FAO, 1973).

Recent reports indicate that the fishing effort is poorly distributed on dermersal stocks of European hake, Mauritanian and Senegalese hake, sea bream species and cephalopods. Previous selectivity studies indicated that the optimum mesh size for hakes was 70 mm and for the other species 60 mm. The tropical zone stocks of shrimp and sea bream are being increasingly harvested.

In regard to pelagic stocks of sardinella, horse mackerel, mackerel, sardines and anchovy, optimum exploitation rates may not have been reached. However, in the southern tropical zone sardinella, mackerel and anchovy are approaching full exploitation.

African governments have expressed concern that in these fisheries the largest part of the catch (66% or nearly 2 million tons) is harvested by fleets not originating from coastal countries, while only 34% is caught by coastal African countries. This contrasts with the position 14 years previously when 86% or 490 000 tons were caught by coastal countries and 14% or 80 000 tons by non-coastal or distant-water fishing.

To establish rational and regulated harvesting would require full collaboration of all 42 nations fishing the resource in the collection of basic data and detailed catch statistics of major species caught. Apart from this there is the problem of evaluating a complex fishery which has seen such a sudden development and has not had a long enough period of sustained and relatively intensive fishing. Many of the effects on stocks of pelagic species are difficult to discern or interpret, and the experience of California, Norway and Peru indicate how problematic and unpredictable such fisheries are, even when subject to the best available scientific scrutiny and assessment (Gulland et al., 1973).

Here the strategy of management of a complex ecosystem under active exploitation has to take into account the need for non-discriminatory regulations that will permit the developing coastal countries to increase their participation in the profitable fisheries off their shores. How is rational management of this ecosystem to result?

Shrimp fisheries. Tropical shrimp fisheries on the continental shelf present a special problem for countries in need of foreign exchange and of fish. Shrimp is a very valuable product while the fish caught as the by-catch of shrimp trawlers often represent 85–90% of the catch by weight, 7 times the bulk of the shrimp caught, but only about one third of the value. The trawlers are designed and managed on an operational basis for shrimp fishing. The aim is to get a gross value of shrimp of more than $100 000 per boat annually. With a crew of four and a hold capacity of 15 tons, the by-catch of fish is jettisoned. This is a very harmful management practice for the ecosystem (Allsopp, 1968). It may be illustrated by the situation off the Guianas (South America) where some 500 shrimp trawlers land annually 23 000 tons of shrimp and jettison approximately 200 000 tons of fish, while the Caribbean region was a net importer of fresh fish, equivalent to 150 000 tons in 1972.

Given the economic obstacles to the operation of fish and shrimp processing on the same vessels, a new strategy has still to be evolved in this and similar tropical shrimp fisheries across the tropical world. An interesting research project is being sponsored by IDRC in Guyana (IDRC, 1973), involving the use of the fish through minced or comminuted fish products for human consumption to develop speciality products from the so-called 'trash' fish. Should this approach be successful, then its application can result in a more rational harvesting of fish resources permitting greater food availability without extra fishing effort.

Coral reef products. Tropical coral reefs have become a source of important fisheries for spiny lobster, *Panulirus sp.* in Belize (Central America) and for novelty shells in the Philippines respectively. These fisheries present special problems of ecosystem management because of the value of the products and selectivity of the operations. The diversification of fishing operations away from spiny lobster, with the development of other fisheries from conch, *Strombus gigas*, and fish of higher value, while introducing a statistical collection system, has served to stabilize the fishery in Belize (FAO, 1968). Philippine investigations into the culture of the various marine gastropods which are collected on a large scale for trinket manufacture are serving to increase the possibility of reef ecosystem conservation in localised areas. The system of culture and leasehold proprietorship may well prove more effective to collectors than the scouring of reefs in search of scarce shells (Development Bank of the Philippines, 1967).

Similar cultivation of marine aquarium fish may lead to better ecosystem conservation, and this aquaculture on a micro-scale for profitable hobby fish may serve as the basis of research, the results from which may be applicable to marine food fish culture systems.

Oyster cultivation. Tropical oysters have evidently not been effectively cultured to large size, mainly because of their almost year-round reproduction. The spatfall and clustered setting of young oysters crowd the substratum with many small oysters. They not only compete for food but provide no space for their full growth potential. If the growth rate that is obtainable in isolation or without fouling can be obtained practically on an industrial scale yields of about 6–10 times may be possible. The oysters must have enough opportunity for ample food and growth to realise their full potential, but their natural reproduction needs to be inhibited. The strategy that is being tested in Sierra Leone, through an IDRC sponsored project, is to try to locate areas where, through salinity, temperatures, currents or other physical or ecological factors, the oysters will *grow* well but will *not reproduce.* If effective growth and industrial-scale production can be obtained, the sanitation and pollution problem can be tackled simultaneously. However, the consumption patterns of oysters in tropical countries show a preference for steamed and dried products and not raw or half-shell oysters, thus avoiding the very considerable problems of oyster sanitary controls in temperate countries.

The resolution of this problem may permit an extensive application of systems for the production of bivalves in vast areas of tropical estuaries and lagoons. Simultaneously the management of such ecosystems will result.

Effectiveness of management approaches

While international commissions serve as the forum for discussion of problems, and represent the only means for international dialogue and unified application of management strategies, there seems to have been evidence of disenchantment by developing tropical countries in the possibility of objective and non-discriminatory adherence to regulations by the more powerful fishing nations. This has been described as their being 'dependent on the benevolence of their powerful competitors which fish the resources off their shores' (Allsopp, 1974).

As regards the harvesting of marine resources, the U.N. Conference on the Law of the Sea may eventually have some very important consequences on the conservation policies and fishing regulations of such coastal countries which cannot now exploit their coastal resources fully, and these policies may result in more effective application of management machinery in such above-mentioned areas as CECAF, continental shelf shrimp fisheries and even reef fisheries.

New management strategies are already emerging. In this regard may be mentioned the requirement by some governments for shrimp trawlers to land minimum quantities of edible fish; the obligatory deposition of catch statistics by foreign trawlers fishing in some West African coastal areas; the requirement for partial landings of the catch of distant-water vessels in coastal ports; and the licensing of foreign flag fishing in extended territorial seas.

Development perspectives

The attainment of objectives of fish for food by culture rather than capture may supervene progressively as the development of tropical territorial seas become imperative. Such culture and husbandry practices will consciously improve ecosystem management, since they will be closed systems for intensified operations. Research inputs, particularly if internationally sponsored, could hasten the trend in this direction.

The U.N. Law of the Sea Conference may have stimulated considerations involving the new dimension of the value and extent of the subjacent seas in which fisheries will play a vital role and which will require the establishment by developing countries of clear management strategies.

The continuous review process of fisheries bio-data collection and refinement of interpretation in tropical fisheries should progressively permit the adoption of improved management strategies and measures while pursuing realistic and rational systems of exploitation of resources for survival. This may involve the utilisation of new species as well, since new technologies may permit their economic and effective processing as food. The availability of more scientific personnel is also likely to be consequential upon the increased interest in the exploitation of national fisheries. Each case may require a different solution, and it is clearly indicated that the larger tropical river basins and the common property marine resources of tropical fisheries may need immediate specific international assistance until the regions concerned can undertake such actions independently. This is in the greatest interest of the effective utilisation of the world's natural resources with the realistic perspective of human survival.

The implications of the mismanagement of tropical ecosystems are grave and can no longer be considered as national or even regional problems. Not only are the obligations of the developed world and humane perspectives and requirements of the developing world now generally recognised as inseparably complementary, but there is the realisation that in the absence of caution, the consequences can only be disastrous for all. It becomes a contest between human discipline worldwide and the ecological imperatives of relentless Nature, whose interactions are in top gear in the tropical world.

Summary

The complexity of management problems in ecosystems involving tropical fisheries are briefly described in relation to the objectives of most tropical countries to develop their fish resources to satisfy food and employment priorities. Such management actions as were previously applied have been based on limited available data, while technical data are accumulating to justify management measures. Some problems considered to be particularly significant are cited as examples deserving international research support. In tropical freshwaters these include the flood-plain fisheries of Africa; the lakes created through hydroelectric impoundments; aquatic weeds; and fishculture. The last offers much potential for extensive application. Among marine problems considered to be of major significance are the fisheries of the East Central Atlantic; continental shelf shrimp fisheries; coral reef fisheries for spiny lobster and shells; and oyster culture. International fishery commissions, though not perfect, are currently felt to be the most effective vehicle for management measures. The development perspectives for tropical countries now assume a wider dimension with the considerations of the U.N. Conference on the Law of the Sea. The need for ecological and management research in such areas should be partly satisfied by international assistance because of the grave consequences of mismanagement.

References

Allsopp, W. H. L. 1968 ms. Optimum utilisation of fish and shrimp caught on the Guiana continental shelf between the Orinoco and the Amazon estuaries.

Allsopp, W. H. L. 1974 ms. African fisheries: their problems and opportunities and their role in the Sahelian famine (to be published by UN Office of Sahelian Relief).
Development Bank of the Philippines. 1967. Report on fishpond financing, Manila.
FAO. 1968. Investigations into marine fishery management research and development policy for spiny lobster fisheries, FAO Report no. TA 2481, Rome.
FAO. 1970. Provisional indicative world plan for agricultural development, chap. 6. The prospects for world fishery development by 1975 and 1985, Rome.
FAO. 1971. Consultation on the conservation of fishery resources and the control of fishing in Africa.
FAO. 1972. Yearbook of fishery statistics: Catches and landings, vol. 34, Rome.
FAO. 1973. Report of the first session of the FAO fishery committee for the Eastern Central Atlantic (CECAF) working party on resources evaluation, Rome, April 1972.
FAO. 1974a. Report of the TAC Working Group on Aquaculture, Rome. February 1974.
FAO. 1974b. Aquaculture Research Programme: Report of TAC Subcommittee on aquaculture, Washington, July 1974.
Gulland, J. A., J. P. Troadec and E. Q. Bayagbona. 1973. Management and development of fisheries in the Eastern Central Atlantic: Technical conference on fishery management and development. J. Fish. Res. Bd. Can. 30(12): 2264–2275.
IDRC. 1973. Project on edible fish products from Caribbean species: Project document of International Development Research Centre, Ottawa Canada.
Manatee workshop. 1974. An international centre for manatee research. Report National Science Research Council, Georgetown, Guyana.
Obeng, L. E. (ed.) 1969. Man-made lakes: The Accra symposium. Ghana University Press, Accra.

Author's address:

W. H. L. Allsopp
International Development Research
Centre Box 8500 Ottawa
c/o University of British Columbia, Duke Hall
Vancouver 8, B.C.
Canada

Man-made natural ecosystems in environmental management and planning

E. van der Maarel

Introduction

This paper has been conceived from a background of European phytosociology and plant ecology, involved in the study of the vegetation of more or less densely populated countries like the Netherlands. I want to speak on behalf of many colleagues from that field of ecology. We are in the awkward position that the changes in the natural environment of our countries are, or at least have been, so tremendously fast that until recently we were spending our time in trying to preserve the remnants of our natural heritage rather than in analysing them. Despite this effort, Europe is losing its biotic diversity at an alarming speed. Unfortunately we have ample circumstantial evidence for this decline, but we need facts and figures to persuade those people who can most effectively try to stop, or at least try to curb, this tendency. Nowadays we consider that the planning authorities are the first people to persuade.

Let me present some of the little factual evidence that we have.

The decline of biotic diversity in some European countries

For various European countries data have been presented on the decrease in the number of native plant species (vascular plants), mostly in comparison with the situation 50–100 years ago.

In Poland various regions have lost 10% of their flora in the last 100 years (Kornas, 1971). In Belgium a loss of 5% is estimated for the whole country, whilst another 5% is considered as threatened with extinction. Even nearly 20% of the bryophytes have disappeared. (Delvosalle et al., 1969). In Great Britain about 8% of the native vascular plant species is extinct or nearly so. For the Netherlands a figure of 4% has been reported for the vascular plants (Westhoff, 1956), whilst for bryophytes and lichens the figures may be well beyond 15% and 30% respectively (cf. Barkman, 1961).

Sukopp (1971, 1972) who presented and discussed many of these data, showed that urban regions have even higher losses, for example 12% in West Berlin which he studied in detail.

From distribution maps on a 20 sq km grid basis for the Netherlands (unpublished sources, Rijksherbarium Leiden) available for the periods around 1900 and 1970 I have estimated losses of vascular plants in a more quantitative way (van der Maarel, 1971). In addition to the number of extinct species, 50 (4%) already quoted from Westhoff, 50 species were found to be extremely rare (i.e. occurring in 1–3 grid units), whilst another 80 species were found to be restricted to only 4–10 grid units. Together these comprise 14% of extinct or very rare species.

In smaller areas even higher figures were calculated, for example the average species number for 20 sq km dropped from 250 to 180, thus nearly 30%. The total number of occurrences in grid units of 606 species now considered extinct or rare, even dropped by 80% between 1900 and 1970.

In this study I also estimated the distribution of species rarity over 19 sociologic-ecological groups. When we take the species groups of all more or less semi-natural environments together, (i.e. species with their optimum, or even entire, development within such environments) we come to over 700 species, or 50% of the Dutch flora, and 60 to 70% of the decline in floristic richness in the Netherlands may be accounted for by losses in the semi-natural categories.

Sukopp (1971) presented similar documentation on the flora of West Berlin. From his data we estimate that half of the 12% species lost were growing in semi-natural habitats.

In conclusion: the flora of temperate Europe, though not very rich in total has had a high regional diversity bound to the fine grained pattern of land use and the large environmental variation, and this flora is rapidly declining, especially its semi-natural components. Although no faunistic data are available in such detail, the general observations are fully parallel.

Ecosystems, as judged from their plant community component, are equally declining in total distribution area. To mention just one example: the Junco-Molinietum, an association of moist hay fields with a large number of characteristic species, once occurred over millions of hectares in the Netherlands. Nowadays only some tens of hectares are preserved with difficulty in nature reserves.

Man-made natural ecosystems intermediate between 'mature natural' and 'cultural' systems

The loss of species and ecosystems all over the earth is perhaps even more pronounced than in temperate Europe alone. Thus in all parts of the world we are losing biotic diversity. In terms of thermodynamics: we are converting biotic negentropy. We should realise, however, that part of this conversion is a by-product of the creation of cultural negentropy, i.e. the building of highly complex urban systems. Man aims principally at creating complexity and stability in order to avoid, or at least to postpone, energy degradation. Here lies the conflict between nature and culture, which we can approach with the help of some general statements. The first one is Norbert Wiener's dictum: 'Life is an island here and now in a dying world'. Man is now trying to extend both the 'here' and the 'now' at the expense of the island's fundaments. The noösphere is certainly the most powerful sphere in influencing spheres from which it evolved, but the noösphere is also the most dependent one, bound to the persistence of zoo-phyto-pedo-litho-hydro-atmo and cosmosphere. Early man was only capable of hunting the zoösphere and exploiting the phytosphere; soon came societies capable of changing the pedosphere and mining the lithosphere. Contemporary society is already taming, and polluting, the hydro- and atmosphere. (Let us hope that our attempts to change cosmospheric conditions will remain primitive for a while since there seems to be no more effective way of destroying our fundament than by changing the cosmosphere.)

The conclusion is obvious: we shall have to find a balance between noöspheric gain and biospheric loss of negentropy.

This brings me to a second statement with a motto paraphrased from a dictum referring to my own country: 'God created the Earth, but man made the natural systems'. This is to say that in Europe—and I think in most of the inhabited world—the natural ecosystems we are said to study and to consider the aim of nature conservation are not natural, except for a very few places with hardly any life at all: they are man-made natural ecosystems. Either man has changed the structure, if he exerts some form of exploitation, or a natural development is taking place on a former non-natural site. So man-made natural systems are in the very focus of environmental management.

Types of man-made natural ecosystems

Let us try some classification—or rather let us do some ordination, with three criteria, since we need at least three axes of differentiation within the very tentative scheme I give you now (Fig. 1).

The first criterion is the rate of influence of man on the biotic component of the ecosystem under influence. In Holland and Germany we used to distinguish roughly four categories, which were applied both to ecosystems and to larger landscape units. In the terms of Westhoff (1952) these are: natural, sub- (or quazi-) natural, semi-natural, and cultivated. A more complicated classification was proposed by von Horstein (quoted by Ellenberg, 1963). Here I combine two classification systems. The first is the classification by Westhoff, introduced by him, and also by Duffey, to the English-speaking world during the 1970 'Symposium on the Scientific Management of Animal and Plant Communities for Conservation' of the British Ecological Society (Westhoff 1971, Duffey 1971). This symposium was largely devoted to semi-natural ecosystems. The second is a recent comprehensive scheme of degrees of human influence by Sukopp (1971) with 'hemerobiotic state' (from hemeros-cultivated, see Jalas, 1965) as a central concept. Table 1 presents an adaptation of this scheme.

We now distinguish six categories: natural, near-natural, semi-natural, agricultural, near-cultural and cultural. We can reasonably couple them to Sukopp's terms hemerobiotic, oligo-, meso, eri-, poly- and metahemerobiotic.

From top to bottom in the scheme (in Table 1) the substrate is increasingly changed, up to the creation of artificial substrates in the cultural ecosystems. Vegetation structure and floristic composition are hardly changed in the natural, to completely extinct in certain cultural systems. Sukopp was also able to check his categories with losses in native species and gains in neophytic species.

The second criterion in our classification is the origin of the ecosystem under study. Did it develop from a former natural situation, for example on a site disturbed by fire through lightning—or did it develop from less natural situations? We may distinguish four categories here: natural, semi-natural, agricultural and of cultural origin.

A third criterion is the present state of development. Let us for the sake of simplicity only distinguish between pioneer, development and mature stages. This alone already gives us 72 categories. So we have to be careful when we speak of man-made or natural systems. Figure 1 indicates the range of man-made natural ecosystems in which we are now especially interested, and which we usually call semi-natural. In general, the developmental stages of succession from abandoned—or created—cultural substrates are included in a wider concept of semi-natural systems. The crucial point in the real semi-natural systems is that their structure has been changed permanently, mostly with a less developed growth form as the dominant layer, whilst the plant and animal species may quite spontaneously enter and leave the system within the limits of their

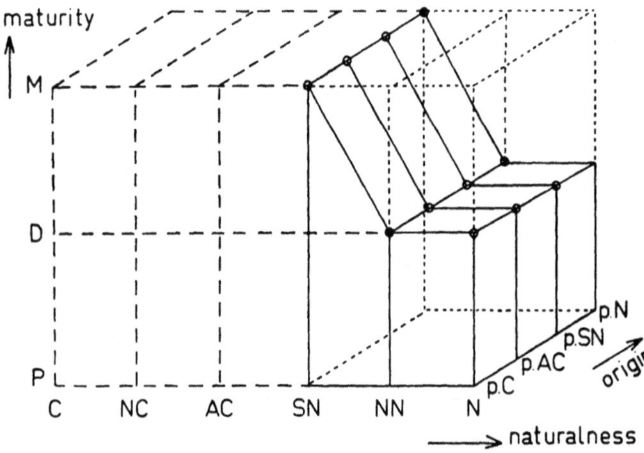

Figure 1. Types of man-made natural ecosystems, indicated in a three dimensional scheme. Axis 1 denotes the present degree of naturalness: C = cultural, NC = near-cultural, AC = agricultural, SN = semi-natural, NN = near-natural, N = natural. The second axis denotes the present degree of maturity: P = pioneer stage, D = development stage, M = maturity stage. The third axis denotes the origin of the ecosystem: pC = post-cultural, pAC = post-agricultural, pSN = postsemi-natural, pN = post-natural. Semi-natural types in a broad sense are indicated with ∅. They include mature stages under semi-natural conditions, and development stages under near-natural and natural conditions.

Table 1. Degrees of naturalness, hemerobiotic state and some characteristics of vegetation and soil, partly after Sukopp (1971).

Naturalness	Hemero-biotic state	Changes substrate	Changes vegetation structure	Changes floristic composition	Loss natives	Gain neo-phytes
natural	a-hemerobiotic	no	no	no	0	0
near-natural	oligo-	few	no	most species spontaneous	<1%	<5%
semi(agri-) natural	meso-	small, superficial	other life form dom.	most species spontaneous	1–5%	5–12%
agricultural	eu-	moderate to drastic	crops dominating	few species spontaneous	>6%	13–20%
near-cultural	poly-	drastic; artificial substrate	open, ephemeral	few to no species	?	21–80%
cultural	meta-	ibid	—	—	—	—

adaptation to the type of management, which is often a more or less regular removal of the vegetation, generally by mowing, burning, cutting sods and extensive grazing.

Three interesting features now emerge:

1. Countries like the Netherlands were once covered by large areas of semi-natural vegetation, such as hay meadows, dune grasslands, rabbit-influenced scrubs, grazed salt marshes, calcareous grasslands and marshes, fens, reed swamps, willow open woodlands coppices. Many plant species are bound to, or at least favoured by, semi-natural environments. In the Netherlands these species constitute 60% of the total flora of 1400 species. Another 20% are bound to, or preferably found in, agricultural and subcultural environments. The remaining 20% are species from woodlands, bogs, fens, lakes, dunes and salt marshes as far as they may be considered near-natural.

2. A second interesting feature is that many of these semi-natural types are relatively old, that is the management has been continued for tens, sometimes for hundreds of years. In relation to the average age of the highest developed life-forms (mostly hemicryptophytes, also chamaephytes, seldom nanophanerophytes) we may even say that they are very old. As a consequence one might expect many of the semi-natural communities to be saturated in the sense that most of the potentially characteristic species have reached the community.

3. A third feature is that due to the relatively fine structure of the vegetation, eventually occurring micropatterns, especially microgradients in the soil, are not levelled by the action of large growth-forms so that very subtle environmental differences may persist and species-rich community complexes may develop. To

mention some figures from own observations: over 130 vascular species in a dune grassland of 2000 sq m, 90 species on 200 sq m in an open dune scrub, 70 species on 2 sq m in a mowed dune slack.

Van der Maarel & Leertouwer (1967) presented information on such a gradient situation in a dune slack bordered by a small elevation. This slack had a connection with the adjoining salt marsh and until recently seawater could reach the spot. The vegetation is low and dense and is heavily grazed by rabbits. Here a height difference of only 16 cm over 10 m has created a distintc pattern of water regimes with the lower parts permanently, and the higher parts mostly only temporarily, inundated in winter. This pattern coincides with a variation in pH from 4.2 in the higher parts to 7.0 in the lower parts. There are three plant communities and a large number of plant species. There is also a clear pattern of species richness, with up to 43 species per sq m in the middle range of the gradient.

Such semi-natural types show many of the features attributed by Odum (1969) to mature ecosystems: viz.: the low net community production, high species diversity, well-organized structure, narrow niche specialization, relatively long life cycles, rather closed mineral cycles and a low level of resilience.

The narrow niche specialisation may be demonstrated with data from Thalen (1971) on part of the same transect, now with squares of only one sixteenth of a sq m, i.e. 25 × 25 cm. The highest species number found on this area is no less than 30. The distribution pattern of *Linum catharticum* and some other species is quite distinct, in a narrow band of only 50–100 cm width. In this 'microcosmos' we can fruitfully apply Whittaker's (1970) ideas on population structure and niche differentiation, together with the environmental relation theory of van Leeuwen (1966, 1973) comprising a theory of requisite environmental variety for each species in relation to the gradient structure of the environment.

Figure 2 presents an environmental gradient system considered as the main basis for species diversity by van Leeuwen (1968) and constructed in a way that is derived from ordination theory (van der Maarel, 1971). Each complex gradient has a diversity potential value, with acid over base rich as No. 1, organic over mineral as No. 2 and dry over wet as No. 3. The combination of gradients may considerably increase differentation potentials, culminating in acid-organic-dry over base-rich-mineral-wet, as we have with acid peaty hummocks in calcareous marshes (especially found in semi-natural alpine meadows.) You should note that no nutrient availability gradient is indicated here. We shall come back to that point but now state that this gradient structure is strictly bound to oligol trophic and mesotrophic conditions.

Now I would like to mention one example of the other type of semi-natura- vegetation, namely developmental stages on culturally determined new substrates. This is taken from a study by Beeftink (1975).

Figure 3 could well be a 'Whittaker diagram'. However, time is on the x-axis. Because of the gradual desalinization a very similar temporal niche differentiation has occurred.

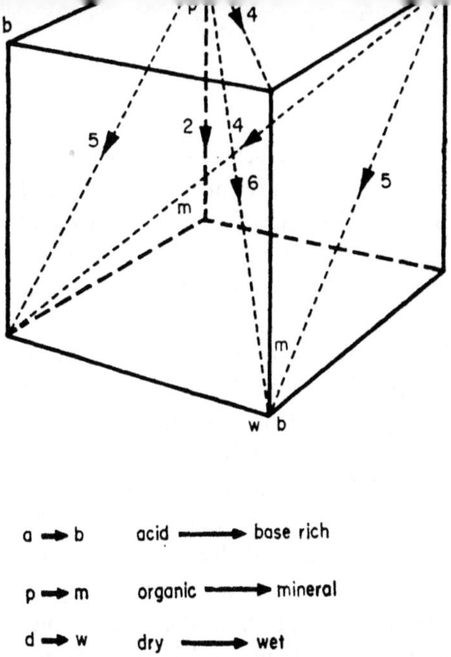

Figure 2. Three dimensional scheme of main complex gradient situations, with indications of the diversity potential value according to van Leeuwen (1968), from van der Maarel (1971).

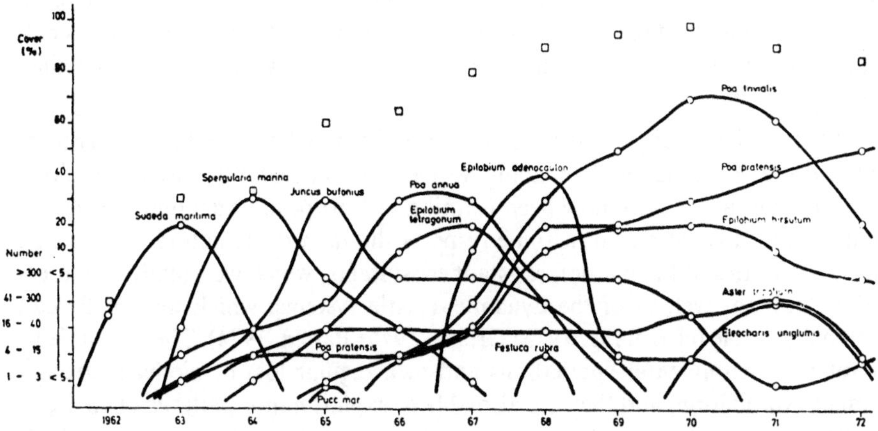

Figure 3. Development of the principal species populations on a permanent plot: a sand flat in the Veerse Meer, an enclosed estuary in the SW Netherlands. Species performance and total vegetation cover (□) are indicated on Doing Kraft's (1954) scale. From Beeftink (1975).

A final underlining of man's positive influence is—I return to the neophytes—the creation of habitats (at least in Europe) in which species new to the regional flora may enter, create their own niches, and in a number of recorded cases may even hybridize directly or through introgression and help to create new species. This is especially interesting in view of the relative poverty of the post-glacial European flora.

These examples concern semi-natural remnants. Indeed any of the micro-gradient environments in Europe are disappearing. Here I come to my third general statement, again with a dictum paraphrased from an American song: 'phosphate is the root of all evil'. I must honestly add the dictum of a German philosopher: 'Ohne Phosphor keine Gedanken', and indeed we cannot think without phosphorus. Increasing evidence is put forward that much of the diversity in semi-natural ecosystems is rapidly destroyed by eutrophication, especially through phosphates. Further, diversity can be diminished by changing or reversing one or more of the gradients such as the three that I have just mentioned, especially by desiccation. Similarly 'diversity in time', i.e. continuous changes in the environment during succession may lead to nothing when these changes are too rapid. To quote Westhoff on the general trend in the agricultural and industrial use of the environment: formerly man did different things at different places but with continuity everywhere; now man does the same thing everywhere but constantly changes the type of activity. To put this into an ecological jargon: the world becomes hyper-ecotonic and badly needs an ecoclinic therapy.

Thus man has been, and is still, able to find subtle balances between natural and cultural demands and in doing so he can considerably enrich his biotic environment, as we have now been learning in European vegetation science for some twenty five years. So in some respects the man-made natural ecosystems share properties of the mature, resistant ecosystems, which we usually call internally stable, as well as of the dynamic, elastic systems, which we usually call externally stable (cf Holling, 1973; Orians, 1975; Patten, 1974). They are able to adapt to environmental fluctuations and to a regular loss of biomass through human exploitation, and they are also able to exploit minor patterns of environmental variation and build up species-rich communities with a relatively high internal stability.

Management of man-made natural ecosystems

At this point we may say something on environmental management, starting with a metaphor derived from the title of an old paper by Westhoff: 'playing chess with nature'. Man plays white—or should we say white man plays? Neither nature nor mankind can afford to get checkmated, so we must reach a draw. And we should play with our pawns rather than with our heavier pieces. That is our fourth statement and it may be specified in some guidelines for environmental management.

guideline 1: Check the gradient structure of the environment and its dynamic properties, whether natural or anthropogenic.
guideline 2: Keep steady-state diverse semi-natural systems in that state by continuing the former management, probably an agricultural one.
guideline 3: Gradients of human influence may be induced in sufficiently large semi- to near-natural areas.
guideline 4: Ecoclines have to be protected from ecotonic disturbance, especially eutrophication, by buffering them.
guideline 5: Ecoclines can be established or amplified by creating oligotrophic conditions dominating over meso- or eutrophic ones.
guideline 6: If a rather constant environment is inevitably getting into a more dynamic state the changes should be damped as much as posssible.
guideline 7: Where natural dynamics occur we should not tame them, and whenever this is inevitable we should replace the natural dynamics by a comparable form of cultural dynamics, for example introducing grazing on a cut-off desalinating natural salt marsh.

These general rules are becoming common practice in various European countries, largely due to the work of van Leeuwen, who was recently awarded the degree of Doctor of Philosophy honoris causa at Groningen University for his very solid contribution to environmental management.

Management and man-made natural systems have so far been mentioned in close connection. I think this is justified, were it merely for the contradiction involved in the term 'management of natural systems'.

Man-made natural ecosystems in environmental planning

Let us end with a fifth statement: 'from pleas to plans'. We have now passed the stage of pleas for preservation. In many countries environmental management is going to be appreciated by physical planning authorities. Environmental management is then carried out on a larger scale and is usually called landscape ecology. Many teams of ecologists, geographers, architects, planners, have recently been formed and involved in providing ecological bases for physical planning. This environmental planning consist of landscape ecological inventories including data-processing; ecological evaluation (i.e. estimation of the functions of the natural environment for human society, for which we are developing a coherent set of criteria); impact analysis (i.e. the study of interactions between functions of the natural environment and society's demands); and finally development of planning models.

In European teams models are mostly in the verbal phase, but on the basis of detailed ecological knowledge we are able to make a reasonable start. In representative studies, like those of the 'Kromme Rijn' area by a team from Utrecht University (Projektgroep Kromme Rijn, 1974, Tjallingii, 1974) and 'Midden-Gelderland' area by Nijmegen University and Provincial Planning Agency of Gelderland (Werkgroep GRAN, 1973, Werkgroep GRIM, 1975), regions of about 500–1500 sq km are carefully analysed with respect to biotic and abiotic components by multidisciplinary teams. Diversity and rarity of plant and selected animal species, of plant communities and soil characteristics, as well as degree of completeness (maturity), are described and taken as evaluation criteria. The resulting evaluation is presented either on a grid basis or on a map showing the basic landscape ecological units. The latter units are usually dstinguished on the basis of the distribution pattern of the potentially natural vegetation. With these evaluation maps the ecologically most significant areas are indicated. In addition, maps of the potential ecological value are presented on the basis of the variety of abiotic, especially soil, characteristics and the spectrum of derivative communities within each type of potentially natural vegetation. Also, as a first outcome of interaction analysis, the sensitivity of ecological values for influences of urbanisation, traffic development etc., through eutrophication, drainage, noise etc. are indicated. Such studies are used by provincial physical planning agencies and may well contribute to better environmental planning.

Landscape ecological studies are now carried out in many European countries, some already on a rather large scale, e.g. in Bavaria, Switzerland, Belgium, Sweden, W. Berlin and various E. European countries. (A survey is published in the Proceedings of the Tagung der Gesellschaft für Ökologie, Erlangen, 1974).

In the Netherlands also a national project is being developed for the National Physical Planning Agency. It is concerned with the building of a general ecological model for the physical planning of the country. It presents a theoretical ecological basis and develops a mathematical interaction model (see van der Maarel, 1975).

It is remarkable that most of the ecological values indicated in these studies refer to man-made natural ecosystems, and that plans for ecological development are largely towards an increase in the total number of possible semi-natural ecosystems within the various potentially natural types. So in large parts of Europe the role of man-made natural ecosystems in environmental planning is very obvious.

Epilogue

Strategies for the management of man-made natural ecosystems must necessarily be of a rather general character. We know far too little of the very complex interaction patterns of man and his natural ecosystems, especially on the landscape level. This brings me to a couple of personal concluding remarks.

Firstly: let us indeed realise that we should not concentrate entirely on model building in relatively simple ecological systems, for the simple reason that by the time our models are powerful enough to treat our complex ecological systems, those systems will have already largely vanished.

Secondly: may I say that landscape ecology might have deserved somewhat more attention during this congress, in view of the urgent need to develop strategies for the management of complex systems and system complexes. May I therefore suggest that INTECOL should encourage further international cooperation in the field of landscape ecology.

References

Barkman, J. J. 1961. De verarming van de cryptogamenflora in ons land gedurende de laatste honderd jaar. Natura 58: 141–151.

Beeftink, W. G. 1975. Vegetationskundliche Dauerquadratforschungen auf periodisch überschwemmten und eingedeichten Salzböden im SW der Niederlanden. In: Symposiumberichte Rinteln 1974. W. Schmidt. ed. Dr. W. Junk, The Hague, J. Cramer, Lehre.

Delvosalle, L., F. Demaret, J. Lambinon and A. Lawalree. 1969. Plantes rares, disparues ou menacés de disparition en Belgique: l'appauvrissement de la flora indigène. Serv. Rés. Nat. dom. Cons. Nat. Trav. 4: 1–128.

Doing Kraft, H. 1954. L'analyse des carrés permanents. Acta Bot. Neerl. 3: 421–424.

Duffey, E. 1971. The management of Woodwalton Fen: a multidisciplinary approach. In: The Scientific Management of Animal and Plant Communities for Conservation. E. Duffey & A. S. Watt. eds. Blackwell, Oxford. 581–597.

Ellenberg, H. 1963. Die Vegetation Mitteleuropas mit den Alpen in kausaler dynamischer und historischer Sicht. Stuttgart.

Holling, C. S. 1973. Resilience and stability of ecological systems. Ann. Rev. Ecol. Systematics 4: 1–23.

Jalas, J. 1965. Hemerobe und hemerochore Pflanzenarten. Ein terminologischer Reformversuch. Acta Soc. Fauna Flora Fenn. 72 (2): 1–15.

Kornas, J. 1971. Changements récents de la flore polonaise. Biol. Conservation 4: 43–47.

Kromme Rijn projekt. 1974. Het kromme Rijnlandschap. Een ekologische visie (with a summary). Stichting Natur en Milieu, Amsterdam.

Leeuwen, C. G. van, 1966. A relation theoretical approach to pattern and process in vegetation. Wentia 15: 25–46.

Leeuwen, C. G. van, 1968. Soortenrijke graslanden en hun milieu. Kruipnieuws 30–1: 16–28.

Leeuwen, C. G. van, 1973. Ecologische systeembeschrijving. In: Oecologie. Diktaat Studium Generale TH, Eindhoven. 165–185.

Maarel, E. van der, 1971. Florastatistieken als bijdrage tot de evaluatie van natuurgebieden (with an English summary). Gorteria 5: 176–188.

Maarel, E. van der, 1975. Naar een globaal ecologisch model voor de ruimtelijke ontwikkeling van Nederland (with an English summary). Rijks Planologische Dienst, Den Haag (in press).

Maarel, E. van der and J. Leertouwer. 1967. Variation in vegetation and species diversity along a local environmental gradient. Acta Bot. Neerl. 16: 211–221.

Odum, E. P. 1969. The strategy of ecosystem development. Science 164: 262–270.

Orians, G. H. 1975. Diversity, stability and maturity in natural ecosystems. In: Unifying concepts in ecology. W. H. van Dobben & R. H. Lowe-McConnell. eds. Dr. W. Junk, The Hague,

Patten, B. C. 1974. The zero state and ecosystem stability. Proc. 1st. Int. Congr. Ecol. Appendix. Pudoc, Wageningen.

Sukopp, H. 1971. Über den Rückgang von Farn- und Blütenpflanzen. In: Belastete Landschaft- Gefährdete Umwelt. G. Olschowy, ed. Goldmann Verlag, München. 364 pp.

Sukopp, H. 1972. Wandel von Flora und Vegetation in Mitteleuropa unter dem Einfluss des Menschen. Ber. Landwirtschaft 50: 112–139.

Thalen, D. C. P. 1971. Variation in some saltmarsh and dune vegetations in the Netherlands with special reference to gradient situations. Acta Bot. Neerl. 20: 327–342.
Tjallingii, S. P. 1974. Unity and diversity in landscape. Landscape Planning 1: 7–34.
Werkgroep Gran. 1973. Biologische kartering en evaluatie van de groene ruimte in het gebied van de stadsgewesten Arnhem en Nijmegen (with an English summary). Rapport Afd. Geobotanie, Nijmegen.
Werkgroep Grim, 1974. Landschapsecologische Basisstudie ten behoeve van het streekplan Midden-Gelderland. Rapport K. U. Nijmegen en P. P. D. Gelderland Arnhem, (in press).
Westhoff, V. 1952. De betekenis van natuurgebieden voor wetenschap en praktijk. Brochure Contactcie. Natuur-en Landschapsbescherming, Amsterdam.
Westhoff, V. 1956. De verarming van flora en vegetatie. In: Vijftig jaar natuurbescherming in Nederland. Amsterdam 151–186.
Westhoff, V. 1971. The dynamic structure of plant communities in relation to the objectives of conservation. In: The Scientific Management of Animal and Plant Communities for Conservation. E. Duffey & A. S. Watt. eds. Blackwell, Oxford 3–14.
Whittaker, R. H. 1970. Communities and Ecosystems. Macmillan, London.

Author's address:

E. van der Maarel
Division of Geobotany
University of Nijmegen
Nijmegen
The Netherlands

Rural ecology and development in Java

Otto Soemarwoto

Introduction

The Government of Indonesia has committed itself to development. From the experience of developed countries and of its own, it has become apparent that development may cause undesirable side effects, e.g. pollution, degradation of natural beauty and also the widening gap between the rich and the poor.

In the Second Five Year Plan, environmental considerations have been included as one of the basic policies. Furthermore in many speeches the President and cabinet members have stressed the need for more emphasis on rural development. Therefore it is timely to study the ecology of rural areas in relation to development.

The rural ecosystem; general description

Figure 1 schematically presents a rural ecosystem. Basically it consists of three subsystems, i.e. the village, the farmfield and the forest. Ecologists have paid much interest to the study of the forest ecosystem. The ecology of the farmfield, of course, has been studied intensively by agriculturists. Recently the attention of ecologists has also been attracted to this ecosystem. The village ecosystem has traditionally been the domain of studies of anthropologists, sociologists and agricultural economists. It has been completely neglected by ecologists. And yet it is a very important part of the rural ecosystem.

An important aspect of the rural ecosystem is the interrelationships of the three subsystems. The right of land of the village does not only extend to the farmfield, but also to part of the forest. This is indicated by the dotted line in Figure 1. This right is invested in the adat-law, i.e. traditional law. It is an unwritten law, but is still observed by the people and the government, regulating a wide variety of rights, such as those concerning land, water, and inheritance. Based on this adat right, the village people exploit the forest for fire and construction wood, rattan, damar and other forest products. In many places in Java, however, the villages are no more in contact with the forest ecosystem, since this latter has been transformed to farmfields to satisfy the need for food of an expanding population.

The farmfield is in constant interaction with the village ecosystem. Solar energy, which has been fixed by the crop-plants, is harvested by the people.

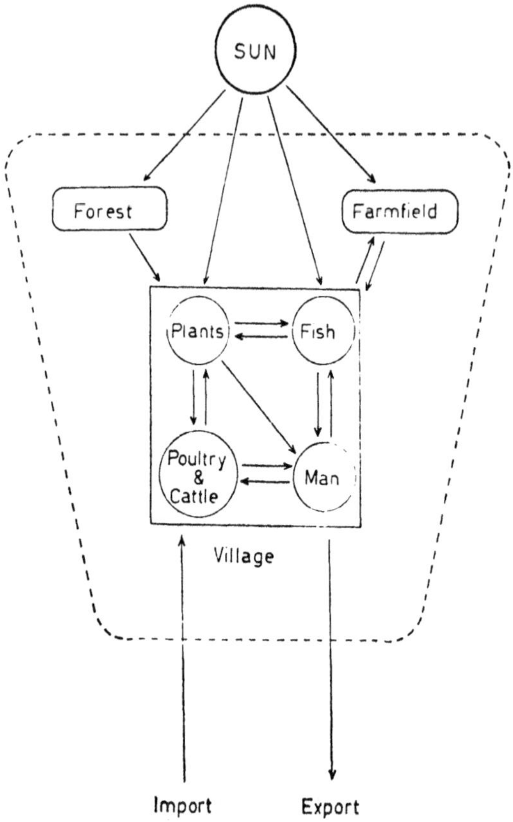

Figure 1. Schematic diagram of a rural ecosystem which consists of three subecosystems, i.e. the village, the farmfield and the forest. Arrows indicate flow of energy and the associated flow of minerals. For further explanation see text. (Size of the three subsystems not drawn to scale.)

Part of this energy is returned to the farmfield in the form of compost. In addition animal and manual energy is invested in the farmfield. In the more modern villages synthetic fertilizers and pesticides also form part of the input in the farmfield.

In discussing the village ecosystem three major aspects deserve attention, i.e. the flow of energy, the cycling of minerals and diversity. In this paper the qualitative nature of these three aspects are described. Further studies are being conducted to obtain quantitative data.

The flow of energy

In rural areas in Java the flow of energy, and associated with it the flow of minerals, is schematically presented in Figure 1.

In addition to providing shelter, the village also plays an important role in the primary production. Travelling through the countryside in Java, one can recognize the existence of a village, not by seeing the houses, but by the group of trees standing amidst the rice fields. In fact it was calculated that the people derived more income from the primary production of the home-gardens than from rice fields (Terra, 1948). Many publications also showed the importance of the home-garden as a source of balanced diet to the people. Many plant varieties also serve as fish and cattle feed.

In West Java it is common for part of the primary production in the home-garden to be channeled into fish, which form a significant source of protein and of income for the people.

Considering the important role of the home-garden in the life of the rural people, extreme care should be exercised in development programs which affects the home-garden.

The kitchen and human wastes are assimilated by fish, cattle and poultry, while the wastes of cattle and poultry are composted and used as fertilizers. Consequently there is little export of energy in the form of waste products.

The more isolated the location of a village, the less are the imports and exports of energy from it. In Java no village is completely isolated from cities. But bad roads in some areas significantly reduce the imports and exports.

The primary source of fuel energy is still wood, although in recent years the use of fossil fuel is increasing. Electricity is only available in very few large villages, those located on main roads and close to cities.

Human and animal energy form the main sources of power in the village ecosystem. Fossil fuel is mostly used for consumptive purposes and little in the production processes. It can be expected that under such conditions the carrying capacity of the environment is low. But the population density of Java is high, with an average figure of 576 persons/km^2 (Iskandar, 1973). In many areas the density exceeds 1000 persons/km^2 (Bappemda, 1972). However, traditional ways of life provide for an equitable distribution of food and income. The difference between the rich and the poor is small.

To increase the carrying capacity of the environment, energy subsidies are required (H. T. Odum, 1971). But still higher levels of energy subsidies are required to improve the living conditions of the people.

In providing energy subsidies to the rural ecosystem care should be taken to ensure that (i) the energy subsidy does not compete with local manual energy, and that (ii) the flow of energy subsidy is as evenly distributed as possible in the ecosystem. The two requirements are closely related to each other.

If the energy subsidy were to be allowed to compete with local manual energy, it would inevitably displace people from their jobs. If allowed to run its own course, the energy subsidy would soon flow through a small segment of the ecosystem. The final result would be more unemployment, the destruction of the traditional food distribution system and the concentration of wealth in a small group of people, who usually came from the cities or even lived in cities. A situation would be created in which one ecosystem, the city, lived at the expense of another ecosystem, the village, and the difference between the city and the village, and between the rich and the poor would become larger. The energy subsidy would not then increase the carrying capacity of the rural environment, and it has rather decreased it, although it did increase the GNP, if computed on a national or regional basis.

A further consequence was that the excess people in the villages were forced either to clear forests on the slopes of mountains for the cultivation of upland crops or to migrate into the cities. In the first instance this causes serious problems of soil erosion and in the second one it further degrades the urban environment.

The problem of rural industrialization is, therefore not only a question of using modern technology versus intermediate technology, but, more important, whether the two requirements mentioned above can be fulfilled. Even small simple machines can and do cause unemployment in villages.

The equitable distribution of the flow of energy subsidy in the rural ecosystem may be achieved either by introducing small machines which are owned by individuals in the villages for small home industries, or by larger factories run on a rural cooperative basis. In either case it is essential that a significant proportion of the energy subsidy should be invested in education to develop the technical capabilities of rural people and their managerial and marketing skill.

With particular reference to energy subsidies in the agricultural sector, the two requirements mentioned above should also be borne in mind, and also that they should not disturb the cycling of minerals and the diversity of the village ecosystem as discussed below. For example, in the use of pesticides and fertilizers most of the energy originates from outside the village ecosystem, i.e. from the fossil fuels as raw materials, transportation of the raw materials to the factories, and from the factories as finished products, energy for production processes, administration, etc., while only very little energy from the village is used. On the other hand when using compost and integrated pest-control much more energy from the village ecosystem itself is used, particularly when natural pesticides are used in the integrated pest-control. This is summarized in Table 1. It is obvious from this table that the use of integrated pest-control with natural pesticides and compost should be much preferred to synthetic pesticides and fertilizers. These latter energy subsidies should be only used as supplementary ones.

The cycling of minerals

As shown in Figure 1 there exists an efficient mechanism for the cycling of minerals in which the people are part of these cycles. In the process the people gain protein and maintain the level of fertility of their farms.

At a first glance the cycle of man → fish → man, which is very common in West Java, seems to be unhygienic. However, the people of Java do not eat raw fish and according to Sri Oemijati, a noted parasitologist at the University of Indonesia, this does not create public health problems (personal communication).

Table 1. Energy subsidy and environmental impact of synthetic pesticides and fertilizers compared to that of integrated pest-control and compost.

Energy subsidy	Nature of investment	Environmental impact
Synthetic pesticides	Capital intensive, mostly imported	High risk
Integrated pest-control with natural pesticides	Labor intensive mostly domestic and local	Low risk
Fertilizers	Capital intensive, mostly imported	High risk
Compost*	Labor intensive, domestic and local	Low risk

* Strictly speaking compost is not an energy subsidy, since it originates from within the ecosystem.

Due to the modernization process, however, this system is considered not proper. More and more lavatories are being built. Unfortunately this has created problems of sewage disposal and eutrophication of surface water, which has caused blooms of aquatic weeds and algae. The new system has also eliminated fish from the cycle, which means the loss of valuable protein.

The introduction of synthetic fertilizers has reduced the need of people to undertake the cumbersome process of composting their solid wastes, except in areas with intensive vegetable growing.

The disturbance of the mineral cycle by the use of fertilizers also makes the rural people more dependent on energy subsidies, which have to be imported from outside the ecosystem. The vulnerability of such dependence on imported energy subsidies became evident during the height of the recent energy crisis, when fertilizers became short in supply. It was obvious that the effort for achieving self-sufficiency in food by increasing energy subsidies, was in fact merely moving the difficulties from dependence on food imports to dependence on imports of energy subsidies. It seems clear, therefore, that efforts should be made to maintain the mineral cycles to reduce as much as possible the dependence on imported energy subsidies and at the same time to develop local sources of energy subsidies. Composting of solid waste should be encouraged, particularly in small units which need only small capital investments.

A further consequence of the disturbance of the mineral cycling is the accumulation of solid wastes in many areas. The problem of waste disposal is further magnified by the population increase. Since funds are short for the construction of waste treatment facilities, it appears logical that the maintenance of the mineral cycling is also very important from the point of view of waste disposal. This is what is actually being attempted in the developed countries under the name of recycling. If the very short and direct cycle of man → fish → man looks repulsive to modern man, it could be lengthened and made less obvious, although, of course, he would have to pay for this extension of the cycle. In addition the longer the food chains, which constitute the cycle, the less efficient they are in terms of the final energy yield to be harvested by man.

Diversity

Coming to a Javanese village one is struck by the many species of plants grown in the home-garden. The plants efficiently occupy the space in different layers, from the surface of the soil to more than 20 meters high. The upper stories are occupied by the crown of coconut trees. Fruit trees, such as mangoes (*Mangifera indica* L.) and rambutan (*Nephelium lappaceum* L.) occupy lower stories and at still lower levels we find cassave, corn, hot peppers and others. Creeping on the ground are such plants as sweet potato, while climbers, such as passion fruit, find their way to the upper stories along the branches of high trees. Thus the village ecosystem resembles a tropical forest.

This diversity gives stability to the village ecosystem. There are no major pest outbreaks and consequently there is little need for pest control. But in the rural areas, where a single species is extensively cultivated, outbreaks of plant pests and diseases have been reported. For example mandarin, in the district of Garut in West Java, is almost completely destroyed by a virus disease.

The diversity also enables the people to harvest something every day, either for their own consumption or to be sold. This gives an economic stability to the people. Unfortunately this stability is being maintained at a low level of standard of living which is deteriorating continuously because of population growth. Hence the need for development of the rural areas. However, the goal of rural development should not merely be an economic one, but, more important, the enhancement of the quality of living of the rural people.

Careful consideration indicates that perhaps the stability is the very cause of undevelopment of the village. Therefore could calculated and planned perturbations be induced to stimulate change and development? But should the development substantially reduce the diversity of the village ecosystem, it would make the rural people too dependent on money-lenders. This would have the effect that the rural people would not benefit from the development, and in fact their conditions may even worsen. Clearly it is not an easy problem. Much more study is needed on the relationships between stability and development.

Summary

Inasmuch as environmental problems in Indonesia primarily stem from poverty and from lack of development, these problems can only be overcome by development to increase the standard of living of the people. Since about 80% of the people live in rural areas, more emphasis should be laid on rural development and due considerations be given to rural ecology to prevent undesirable side effects arising in the development process.

Some characteristics of the rural ecology are:

Energy flow. There is more or less equal distribution of energy and consequently also an equitable distribution of income. The introduction of energy subsidies to the village ecosystem should take into account that i) the energy subsidy should not compete with manual energy and ii) the flow of the energy subsidy should be as evenly distributed as possible in the ecosystem in order to prevent the accumulation of power and wealth in a small segment of the ecosystem.

Cycling of matter. Man is part of the food web and is directly involved in the cycling of matter, which is an efficient method of waste disposal and protein production. Rural development programs should endeavour to keep the cycling of matter intact, while improving the hygienic conditions.

Diversity of the ecosystem. The home garden is planted with many species of plants which efficiently occupy the space in several layers. The people also obtain their living from diverse sources. These diversities give stability to the village ecosystem. Economic development should as far as possible maintain this diversity.

References

Bappemda, Djabar. 1972. Facts and figures of West Java. Preliminary. (Indonesian).

Iskandar, N. 1973. Population problem and the human environment. In: Management of human environment and national development. Institute of Ecology and Obor Foundation, Bandung and Jakarta. Indonesian with English summary, p. 75–95.

Odum, H. T. 1971. Environment, power and society. Wiley Inter-Science, New York, London, Sydney, Toronto, 331 pp.

Terra, G. J. A. 1948. Tuinbouw. In: C. J. J. van Hall and G. van de Koppel (Ed.): Landbouw in the Indische Archipel vol. II A, p. 622–746. Van Hoeve, The Hague, Holland (Dutch.)

Author's address:

Otto Soemarwoto
Institute of Ecology
Padjadjaran University
Bandung
Indonesia

Discussion

Summarized by P. Gruys

Participants: the authors W. H. L. Allsopp (Canada), C. S. Holling (Canada), E. van der Maarel (The Netherlands), the chairman W. B. Banage (Zambia), together with E. C. Evans (U.S.A.), M. J. Frissel (The Netherlands), G. Haase (D. R. Germany), A. D. Hasler (U.S.A.), A. M. A. Imevbore (Nigeria), S. Kohlemainen (Puerto Rico), Le Chi Thangh (Vietnam), D. Müller-Dombois (U.S.A.), N. Polunin (Switzerland), G. S. Puri (U.K.), J. E. Satchell (U.K.), Ch. Souchon (France), C. O. Tamm (Sweden), I. S. Zonneveld (The Netherlands).

In many of the mathematical studies presented at this Congress, criteria for stability and optimalisation procedures have been studied by analytical methods. Unfortunately, these methods can only be applied on rather simple models. In dynamic models of real systems, however, hundreds of different equations and functions have to be included. Do you not think the mathematics as they were presented here are insufficient to describe the complexity of ecological systems? If so, do you not think it worthwhile to pay more attention to modern simulation languages? (*Frissel* to *Holling*).

I agree. The information collected in one of the most comprehensive studies of a natural system, the study of the spruce budworm by R. F. Morris and co-workers, has been used to implement several kinds of models for simulation and optimization. None of these models proved to be satisfactory in predicting the behaviour of the system. The best results were obtained when models were used with a constant reconciliation to the real system, in a way of trial and error. Modern simulation techniques can be helpful to this. (*Holling* to *Frissel*).

The consideration of long-term stability, by Holling, brings in an evolutionary dimension. Regional stress factors should be assessed as a key to long-term stability of regional biota. The native Hawaiian biota, presented as an example, appears to be rejuvenated by volcanic perturbations. It is in part better equipped to invade and establish itself on new lava flows than is the exotic biota, which otherwise represents a threat to the native biota in geologically old habitats. Thus volcanism, an evolutionary stress factor in Hawaii, preserves a certain stability of the island biota. (*Müller-Dombois* to *Holling*).

The recent man-made lakes in tropical Africa are too large. The excess storage of water causes management problems. Moreover, valuable land is wasted and the stagnation of the water leads to excessive evaporation and deleterious effects downstream. The possibility of having a series of small dams instead of one large dam, particularly in tropical areas, must be studied. INTECOL might direct its attention to joint research by ecologists and engineers. (*Imevbore* to *Allsopp*).

This suggestion has considerable merit. Generally I think that the site of dams is selected to enable the cheapest feasible construction in relation to the hydroelectric power required to be generated, with less regard for the surface being covered. With newer tropical man-made lakes the socio-economic aspects are being given consideration. The broader ecological aspects also deserve more attention and perhaps INTECOL can use its voice to exert some influence. (*Allsopp* to *Imevbore*).

The chemical control of aquatic weeds is ecological nonsense. Macrophytes must be viewed as a harvestable crop with high quality protein. The organic matter and nutrients should be returned to the land for use as compost or cattle food, and not be allowed to escape into a sink. Do you envisage the use of harvesting machinery for the tropical waters? (*Hasler* to *Allsopp*).

I agree that, as I stressed in my paper, the use of chemicals for weed control is unacceptable. Conversion into comestible products is desirable, if possible. In the tropics, the rate of growth of aquatic vegetation and the sheer mass and extent of the weed infestation require much effort if systems for weed conversion are to have wider or effective application. If harvesting machinery can be effective and economical it will be widely adopted. But would it not consume too much fossil energy? (*Allsopp* to *Hasler*).

Herbicides also come from oil. The trade-off is negative for chemical weed control, because the chemical is wasted, while mechanically harvested weeds can be used (*Hasler* to *Allsopp*).

In addition to the vegetation categories described by van der Maarel, original, natural (spontaneously developed after disturbance of the original vegetation), and artificial (not having evolved naturally in a region) vegetation should be distinguished. The distinction of the latter type is particularly important with respect to the effect of stress factors, which may also be alien to the region concerned. (*Müller-Dombois* to *van der Maarel*).

I agree. In my scheme, original vegetation is called N-p. N-M, i.e. mature, post-natural natural vegetation. Your artificial vegetation is equivalent to the cultural type in my system. (*van der Maarel* to *Müller-Dombois*).

In ecosystem management, the word ecosystem should be understood in a broad sense, including all relevant natural and socio-economic structures and processes. The term 'geosystems' has been used for this concept. (*Haase* to *van der Maarel*).

Good physical planning should take the natural, agricultural and urban ecosystems, as well as society itself, into account. This requires a team of scientists. (*van der Maarel* to *Haase*).

Several questioners presented general comments on the relations between society, politics and ecology.

Man seriously affects ecosystems by war as, in our time, in Angola, Mozambique, Guinea Bissau, some arab countries, and especially in Vietnam, Cambodia and Laos. In Vietnam, millions of tons of napalm, phosphorous and other bombs have been dropped, millions of litres of toxic chemicals and poison gas have been used, and millions of hectares of forest have been defoliated and destroyed. This destruction has been finished by erosion due to heavy tropical rains. Millions of humans have been intoxicated and killed. This will affect the life of our people for decades. We call upon you, scientists, to raise the voice of conscience to stop these wars, and to demand that the U.S. put an end to their military aid to the Saigon administration. We call upon you to help us heal the war wounds, to help us with means and methods to study the ecology and restore its normal functioning in Vietnam. (*Le Chi Thangh*).

What is the position of ecologists in society? Should they blindly take the commands of the authorities, whatever the latters' intents? If so, ecologists might even be exploited to detect the weak spots of ecosystems for the latters' most effective destruction during military action. In fact, ecologists should not only conduct scientific research, but also criticize governments and society if these act against the ecological principles, and provide information to the general public. (*Souchon*).

Political disorder as a factor meriting serious attention when planning strategies for management of ecosystems, was also stressed by *Polunin*.

Very long-term planning is less acceptable to voters than short or mid-term planning. Planning to contain ecological catastrophes, which must be long-term, might therefore be unrealisable in democratic societies. (*Satchell*).

However, long-term planning seems to be difficult under absolutist political systems too. (*Holling*).

Other representatives from democratic countries held a less pessimistic view than Satchell; the procedure for long-term planning adopted in the Netherlands was explained. (*Müller-Dombois*; *van der Maarel*).

To obtain a reciprocal relationship between society and ecology, ecologists should convey their information to the general public in a suitable form. This serves at least two ends, viz. it prevents ecological decisions being taken by ill-informed people, and it affects decision-making politicians through the general public. (*Tamm*; *Kohlemainen*).

Ecologists very often have a negative attitude towards projects aimed at the development of natural resources and the management strategies proposed to this end, which hinders communication with decision-makers. (*Mukiibi*).

Politicians may be convinced by ecological lessons from the past. There are many examples of ecological successes and mistakes, illustrating not only ecological principles but also sociological and economic consequences of interference with the functioning of ecosystems, which could and should be used to guide the decision makers. (*Evans*).

Land appraisal could aid in decision making in development projects. This is a land evaluation technique being developed by soil scientists, to which ecologists should also contribute. (*Zonneveld*).

The study of temperate ecology is far ahead of tropical ecology. Development of the latter, however, is indispensable as a basis for management strategies for tropical ecosystems. Mistakes from the past, such as undue emphasis on economics as the index of development, should be avoided. In the management of tropical ecosystems, the emphasis should be on environmental quality. (*Puri*).

Author index

Subject index

GPSR Compliance

The European Union's (EU) General Product Safety Regulation (GPSR) is a set of rules that requires consumer products to be safe and our obligations to ensure this.

If you have any concerns about our products, you can contact us on ProductSafety@springernature.com

In case Publisher is established outside the EU, the EU authorized representative is:

Springer Nature Customer Service Center GmbH
Europaplatz 3
69115 Heidelberg, Germany

Batch number: 10584442

Printed by Printforce, the Netherlands